LATENT CLASS AND LATENT TRANSITION ANALYSIS

LATENT CLASS AND LATENT TRANSITION ANALYSIS

With Applications in the Social, Behavioral, and Health Sciences

Linda M. Collins
Stephanie T. Lanza
The Pennsylvania State University

WILEY

A JOHN WILEY & SONS, INC., PUBLICATION

Published by John Wiley & Sons, Inc., Hoboken, New Jersey.
Published simultaneously in Canada.

For general information on our other products and services or for technical support, please contact our Customer Care Department within the United States at (800) 762-2974, outside the United States at (317) 572-3993 or fax (317) 572-4002.

Wiley also publishes its books in a variety of electronic formats. Some content that appears in print may not be available in electronic format. For information about Wiley products, visit our web site at www.wiley.com.

Library of Congress Cataloging-in-Publication Data:

Collins, Linda M.
   Latent class and latent transition analysis / Linda M. Collins, Stephanie T. Lanza.
      p. cm.
   Includes bibliographical references and index.
ISBN 978-0-470-22839-5 (cloth)
   1. Latent structure analysis. 2. Latent variables. 3. Statistical methods. I. Lanza, Stephanie T., 1969– II. Title.
   QA278.6.C65 2010
   519.5'35—dc22   2009025970

10 9 8 7

For David and Kathy
with admiration
(LMC)

For Dad
my first math teacher
(STL)

CONTENTS

PART II   ADVANCED LCA

5   Multiple-group LCA   113

List of Figures

List of Tables

xxxLIST OF TABLES

Acknowledgments

Many wonderful people helped us as we wrote this book. We have been fortunate enough to teach several graduate courses at Penn State and workshops at Penn State and other locations with truly outstanding groups of participants. The attendees of these classes and workshops asked many thoughtful and stimulating questions. These questions always prompted us to think more deeply about the material, and have shaped this book in subtle and profound ways. Many of the attendees read draft versions of the book and pointed out errors and places where the exposition could be improved. We would like to thank Dr. Brian Francis of the University of Lancaster, England, and Dr. Maria Rohlinger of the University of Cologne, Germany, for inviting us to give multiple-day workshops where we had the opportunity to learn a bit about the fascinating research questions and issues on the minds of the attendees. We also have benefitted from the questions many scientists have submitted to The Methodology Center's Help Desk over the years. These questions have helped us to stay in touch with the issues that end users face when they apply latent class analysis in their work.

It is impossible to offer enough thanks to our colleagues and trainees at The Methodology Center, or to overstate their influence on this book and on all of our work. From water cooler conversations to email exchanges to interest group meetings to formal presentations, the Center is the scene of an ongoing, lively, interdisciplinary, multi-way conversation about any and all aspects of quantitative methodology. We also want to thank our colleagues and trainees at Penn State's Prevention Research Center, with whom we frequently collaborate. This book is much richer because we have the privilege of working in such a stimulating, challenging, and supportive environment. That said, we note that any errors are our responsibility and ours alone.

We are grateful to the two primary sources of support for the infrastructure of The Methodology Center, namely Penn State's College of Health and Human Development, and the National Institute on Drug Abuse (NIDA). In addition to this infrastructure support, NIDA, particularly the Division of Epidemiology, Services, and Prevention Research (DESPR), also has for many years provided support for our research and that of our colleagues in The Methodology Center. This funding has enabled us to gain whatever expertise we have in latent class and latent transition analysis. Having funding from NIDA and interacting with the outstanding program officials in DESPR has helped to keep us focused on the perspective of social, behavioral, and health scientists.

Several people gave us help with preparation of the manuscript. Our resident experts on LaTeX, Dr. John Dziak and Dr. Bethany Bray (now at Virginia Tech) were extremely generous with their time to us, two LaTeX novices learning the hard way. Dr. YoungKyoung Min and Ms. Lisa Litz helped with checking the manuscript for internal consistency. Ms. Tina Meyers helped in too many ways to enumerate. Ms. (soon to be Dr.) Jessica Johnson was kind enough to help us with last-minute LaTeX formatting during a time crunch, thereby saving our sanity.

Finally, we want to thank John, Matthew, and Joy Graham, and Colton, Sadie, and Maura Williams, who were kind enough to tolerate our preoccupation with this book during the various stages of its preparation.

L. M. C. and S. T. L.

Acronyms

AIC	Akaike information criterion
BIC	Bayesian information criterion
DA	data augmentation
EM	expectation-maximization
GPA	grade-point average
LCA	latent class analysis
LTA	latent transition analysis
MAR	missing at random
MCAR	missing completely at random
MNAR	missing not at random
ML	maximum likelihood
NLSY	National Longitudinal Survey of Youth
RMLCA	repeated-measures latent class analysis
STI	sexually transmitted infection

PART I

FUNDAMENTALS

CHAPTER 1

GENERAL INTRODUCTION

1.1 OVERVIEW

This chapter provides a general introduction to the book, to latent class analysis (LCA) (e.g., Goodman, 1974a, 1974b; Lazarsfeld and Henry, 1968), and to a special version of LCA for longitudinal data, latent transition analysis (LTA) (e.g., Bye and Schechter, 1986; Langeheine, 1988). (Unless we indicate otherwise, when we discuss the latent class model in general we are referring to both LCA and LTA.) We discuss the conceptual foundation of the latent class model and show how the latent class model relates to other latent variable models. Two empirical examples are presented, both based on data on adolescent delinquency: one LCA and the other LTA. These empirical examples are discussed in very conceptual terms, with the objective of helping the reader to gain an initial feeling for these models rather than to convey any technical information. Next is an overview of the remaining chapters in the book. This chapter ends with some information about sources of empirical data used for the book's examples, information about software that can be used for LCA and LTA, and a discussion of the additional resources that can be found on the book's web site.

Latent Class and Latent Transition Analysis. By Linda M. Collins and Stephanie T. Lanza
Copyright © 2010 John Wiley & Sons, Inc.

1.2 CONCEPTUAL FOUNDATION AND BRIEF HISTORY OF THE LATENT CLASS MODEL

Some phenomena in the social, behavioral, and health sciences can be represented by a model in which there are distinct subgroups, types, or categories of individuals. Many examples can be found in the scientific literature. One example is Coffman, Patrick, Palen, Rhodes, and Ventura (2007), who identified subgroups of U.S. high school seniors who had different motivations for drinking. Another example is Kessler, Stein, and Berglund (1998). Based on a sample of U.S. residents between the ages of 15 and 54 who participated in the National Comorbidity Survey (Kessler at al., 1994), Kessler et al. identified two types of social phobias. A third example is Bulik, Sullivan, and Kendler (2000), who identified six different categories of disordered eating in a sample of female twins, also U.S. residents. Each of these studies used LCA to identify subgroups in empirical data.

As the name implies, LCA is a latent variable model. Readers may be acquainted with other latent variable models: for example, factor analysis. (How LCA relates to other latent variable models is discussed in Section 1.2.1.) The term *latent* means that an error-free latent variable is postulated. The latent variable is not measured directly. Instead, it is measured indirectly by means of two or more observed variables. Unlike the latent variable, the observed variables are subject to error. Most statistical analysis approaches based on latent variable models attempt to separate the latent variable and measurement error.

The scientific literature has used a variety of terms for latent variables and observed variables. Latent variables are often referred to as *constructs*, particularly in psychology and related fields (Pedhazur and Schmelkin, 1991). In this book we sometimes refer to the observed variables as *indicators* of the latent variable, to emphasize their role in measurement. We also use the term *item* when we are referring to particular questions on data collection instruments such as questionnaires or interviews.

Figure 1.1 illustrates a hypothetical latent variable. In the figure the latent variable is represented by an oval. The observed indicator variables measuring the latent variable are represented by squares labeled X_1, X_2, and X_3. The circles containing the letters e_1, e_2, and e_3 represent the error components associated with X_1, X_2, and X_3, respectively. There are arrows running from the latent variable to each indicator variable, as well as arrows running from each error component to each indicator variable. These arrows represent an important concept underlying all latent variable models, including LCA: The causes of the observed indicator variables are the latent variable and error. It is particularly noteworthy that the causal flow is *from* the latent variable *to* the indicator variable, not the other way around. That is, observed indicator variables measure latent variables, but the observed indicator variables do not cause the latent variables.

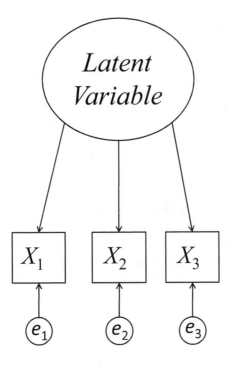

Figure 1.1   Latent variable with three observed variables as indicators.

In LCA each latent variable is categorical, comprised of a set of latent classes. These latent classes are measured by observed indicators. In Coffman et al. (2007) the latent variable was motivation for drinking. The latent classes consisted of one group of high school seniors motivated primarily by wanting to experiment with alcohol; a second group made up of thrill-seekers; a third group motivated primarily by the desire to relax; and a fourth group motivated by all of these reasons. Coffman et al. measured motivations for drinking using questionnaire item data from Monitoring the Future (Johnston, Bachman, and Schulenberg, 2005). In Kessler et al. (1998) the latent variable was social phobia. The latent classes were those with fears that were primarily about speaking, and those with a broader range of fears. Kessler et al. measured social phobia using interview data from the National Comorbidity Survey (Kessler et al., 1994). In Bulik et al. (2000) the latent variable was disordered eating, consisting of the following six latent classes: Shape/Weight Preoccupied; Low Weight with Binging; Low Weight Without Binging; Anorexic; Bulimic; and Binge Eating. Bulik et al. measured disordered eating based on symptoms obtained from detailed interviews.

1.2.1   LCA and other latent variable models

A number of latent variable models are in wide use in the social, behavioral, and health sciences (e.g., Bollen, 1989, 2002; Bollen and Curran, 2005; Jöreskog and Sörbom, 1979; Klein, 2004; Nagin, 2005; Skrondal and Rebe-Hesketh, 2004; Von Eye and Clogg, 1994). One of the best-known is factor analysis (e.g., Gorsuch, 1983; McDonald, 1985; Thurstone, 1954). The latent class model is directly analogous to the factor analysis model. Both models posit an underlying latent variable that is measured by observed variables. The key difference between the latent class and factor analysis models lies in the nature and distribution of the latent variable. As mentioned above, in LCA the latent variable is categorical. This categorical latent variable has a multinomial distribution. By contrast, in classic factor analysis the latent variable is continuous, sometimes referred to as *dimensional* (Ruscio and Ruscio, 2008), and normally distributed. Ruscio and Ruscio (2008) define *categorical latent variables* as those in which "qualitative differences exist between groups of people or objects" and *continuous* (or *dimensional*) *latent variables* as those in which "people or objects differ quantitatively along one or more continua" (p. 203). In both LCA and factor analysis, the observed variables are a function of the latent variable and error, although the exact function differs in the two models. To date, considerable work has been done concerning continuous latent variables (e.g., Bollen, 1989; Jöreskog and Sörbom, 1979; Klein, 2004). There has been somewhat less research on categorical latent variable models, but interest in this topic appears to be growing.

Table 1.1 shows how LCA relates to some other latent variable models for cross-sectional data. As Table 1.1 shows, latent variable models can be organized according to (a) whether the latent variable is categorical or continuous, and (b) whether the indicator variables are treated as categorical or continuous. Sometimes the distinctions between the various models are a bit arbitrary, but we make them, nevertheless, to help clarify where latent class models fit in with other latent variable models and to help illustrate what kinds of models we discuss in this book. Models in which the latent variable is continuous and the indicators are treated as continuous are referred to as *factor analysis*. When the latent variable is continuous and the indicators are treated as categorical, this is referred to as *latent trait analysis* or, alternatively, *item response theory* (e.g., Baker and Kim, 2004; Embretson and Reise, 2000; Langeheine and Rost, 1988; Lord, 1980; Van der Linden and Hambleton, 1997). Approaches in which the latent variable is categorical and the indicators are treated as continuous are generally referred to as *latent profile analysis* (e.g., Gibson, 1959; Moustaki, 1996; Vermunt and Magidson, 2002), although they are sometimes referred to as latent class models. In this book, when we refer to latent class models we mean models in which the latent variable is categorical and the indicators are treated as categorical.

Table 1.1 is intended as an overview rather than a complete taxonomy of all latent variable models. Therefore, it does not mention all latent variable models. For ex-

Table 1.1   Four Different Latent Variable Models

	Continuous Latent Variable	Categorical Latent Variable
Indicators treated as continuous	Factor analysis	Latent profile analysis
Indicators treated as categorical	Latent trait analysis or item response theory	Latent class analysis

ample, there are latent variable models that treat the indicators as ordered categorical, count data, or other metrics (e.g., Böckenholt, 2001; Vermunt and Magidson, 2000).

1.2.2   Some historical milestones in LCA

In this section we briefly present some historical milestones in LCA. This section is intended not to be a comprehensive account of important work in LCA, merely to note some work that is particularly relevant to this book. Thus much important work is necessarily omitted. More detailed histories of LCA may be found in Goodman (2002), Langeheine (1988), and Clogg (1995).

One early major work on latent class analysis was the book by Lazarsfeld and Henry (1968). They were not the first to suggest the idea of a categorical latent variable, but their book represents the first comprehensive and detailed conceptual and mathematical treatment of the topic. Although Lazarsfeld and Henry convincingly demonstrated the potential of latent class analysis in the social and behavioral sciences, the lack of a general and reliable method for obtaining parameter estimates was a major barrier that prevented widespread implementation of their ideas for a time.

This changed in the next decade when Goodman (1974a, 1974b) developed a straightforward and readily implementable method for obtaining maximum likelihood estimates of latent class model parameters. Goodman's approach to estimation was later shown to be closely related to the expectation-maximization (EM) algorithm (Dempster, Laird, and Rubin, 1977). The EM algorithm is used in much LCA software today.

The latent class model was made much more general when it was placed within the framework of log-linear models (Formann, 1982, 1985; Haberman, 1974, 1979; Hagenaars, 1998). This opened up a number of possibilities for model fitting (e.g., Rindskopf, 1984) and set the stage for some important new developments. One development was the incorporation of covariates into latent class models (Dayton and Macready, 1988). Another was models that identify latent classes based on individual growth trajectories in longitudinal data (e.g., Muthén and Shedden, 1999; Nagin, 2005).

A variety of approaches have been developed for modeling changes over time in latent class membership. Many of these have fallen in the general family of Markov models (Everitt, 2006). In these models a transition probability matrix represents the probabilities of latent class membership at Time t conditioned on latent class membership at an earlier time, Time $t-1$. Early examples of work in this area include Bye and Schechter (1986), Langeheine (1988), and Collins and Wugalter (1992).

1.2.3   LCA as a person-oriented approach

Bergman and Magnusson (1997; Bergman, Magnusson, and El-Khouri, 2003) have drawn a distinction between variable-oriented and person-oriented approaches to statistical analysis of empirical data in the social and behavioral sciences. In *variable-oriented approaches* the emphasis is on identifying relations between variables, and it is assumed that these relations apply across all people. Traditional factor analysis is an example of a variable-centered approach. The emphasis in factor analysis is on identifying a factor structure that accounts for the linear relations among a set of observed variables. The factor structure is assumed to hold for all individuals. In contrast, in *person-oriented approaches* the emphasis is on the individual as a whole. As Bergman and Magnusson stated: "Operationally, this focus often involves studying individuals on the basis of their patterns of individual characteristics that are relevant for the problem under consideration" (p. 293). At the same time, as Bergman and Magnusson pointed out, the focus of most scientific endeavors is nomothetic; that is, the goal is not merely to study individuals, but to reason inductively to draw broad conclusions and identify general laws. One way to do this in a person-oriented framework is to look for subtypes of individuals that exhibit similar patterns of individual characteristics. LCA does exactly this, and therefore is usually considered a person-oriented approach.

1.3   WHY SELECT A CATEGORICAL LATENT VARIABLE APPROACH?

Why select a model that posits a categorical latent variable, like LCA, instead of a model that posits a continuous latent variable, like factor analysis? One reason is to identify an organizing principle for a complex array of empirical categorical data. As will be seen in the empirical examples throughout this book, with LCA an investigator can use an array of observed variables representing characteristics, behaviors, symptoms, or the like as the basis for organizing people into two or more meaningful homogeneous subgroups. The array of observed data is usually much too large and complex for the subgroups to be evident from inspection, even very painstaking inspection, alone.

Another reason for selecting a model that posits a categorial latent variable might be that an investigator believes that a particular phenomenon is inherently categorical and therefore must be modeled this way. If research questions revolve around whether a phenomenon is in some sense "truly" categorical or continuous, it may be helpful to explore this empirically. This issue is complex and has been explored at length elsewhere. We refer the interested reader to the taxometric method developed by Meehl (1992) and described in detail by Ruscio, Haslam, and Ruscio (2006; Ruscio and Ruscio, 2008).

Although debating whether a particular phenomenon is inherently continuous or categorical can be fascinating, in this book we remain in general agnostic about this issue, while recognizing that in many specific situations it is important. Our view is that many phenomena may have both continuous and categorical characteristics. As an example, consider alcohol use. Alcohol use can be considered a continuous phenomenon; for example, it may be operationally defined as the number of ounces of alcohol consumed per day. Alternatively, it may be considered a categorical phenomenon; for example, it may be operationalized as categories such as non-use, social use, dependence, and abuse. Rather than debate whether alcohol use is truly continuous or categorical, we would rather consider whether a continuous or categorical operationalization of alcohol use is more relevant to the research questions at hand. Our perspective is that any statistical model, including the models presented here, are lenses that can be used by investigators to examine empirical data. The worth of such a lens lies in the extent to which it reveals something both interesting and scientifically valid. Therefore, in this book we do not take a stand on whether the latent variables we are modeling are in some sense truly categorical. Instead, we merely make the modest assumption that given a particular array of research questions and the empirical data set at hand, a categorical model can provide some useful insights.

1.4   SCOPE OF THIS BOOK

This book is intended to provide the reader with an advanced introduction to LCA and LTA rather than to provide a comprehensive review of the literature on latent class models. Therefore, we have limited the scope of the book to LCA and LTA with categorical indicators. This means that we have had to leave out some important and interesting areas. One such area is latent profile analysis, mentioned briefly above (e.g., Böckenholt, 2001; Vermunt and Madigson, 2002). Other topics not covered in this book include multilevel LCA and LTA (e.g., Asparouhov and Muthén, 2008; Vermunt, 2003), LCA involving complex survey data (e.g., Patterson, Dayton, and Graubard, 2002), and models for latent classes of growth curves (e.g., Muthén and Shedden, 1999; Nagin, 2005). LCA and LTA are part of a larger group of models known in statistics as *mixture models* (e.g., McLachlan and Peel, 2000). We do not cover the many variations of mixture models, such as factor mixture models (e.g., Lubke and Muthén, 2005). However, we do provide readers with some background that may be helpful if they read the literature in these and other areas in LCA and related fields.

Latent class models have been expressed in the literature in two different ways. One way has been in probability terms, following the general approach of Goodman (1974a, 1974b; see also Rubin and Stern, 1994). The other way has been in log-linear model terms, following the general approach of Haberman (1979) and Formann (1982, 1985). In this book we use probability terms, because we find this a natural vehicle for explaining and understanding latent class models. However, most latent class models can be expressed either way.

1.5   EMPIRICAL EXAMPLE OF LCA: ADOLESCENT DELINQUENCY

To provide a first exposure to LCA and to give a flavor for the kinds of statistical models we discuss throughout the book, in this section we present a latent class model of adolescent delinquent behavior. We suggest that, for now, readers not concern themselves with any of the technical details. These are covered in later chapters. Instead, we hope that these examples will provide a conceptual feel for the kinds of research questions that can be addressed using LCA.

The analyses presented here are based on $N = 2,087$ adolescents who participated in the National Longitudinal Study of Adolescent Health (Add Health) study (Udry, 2003; see Section 1.8.1.1) and were part of the public-use data set. The subjects were in Grades 10 and 11 (mean age = 16.4) in the 1994–1995 academic year and provided data on at least one variable measuring adolescent delinquency at Wave I. Among the data collected on these adolescents were responses to questionnaire items about delinquent behavior. In the analyses reported in this book we used questions that asked whether the student had ever engaged in the following behaviors: lied to their parents about where or with whom they were; acted loud, rowdy, or unruly in public; damaged property; stolen something from a store; stolen something worth less than $50; and taken part in a group fight. The students could choose one of the following response options: "Never," "1–2 times," "3–4 times," or "5 or more times." Responses were recoded to "No," meaning that the student had not engaged in the behavior in the past year, or "Yes," meaning that the student had engaged in the behavior one or more times in the past year. The proportions of recoded "Yes" responses to each question are shown in Table 1.2.

The data in Table 1.2 show that some of the delinquent behaviors are more normative than others. For example, it is not uncommon for an adolescent to lie to his or her parents about whereabouts and companions. Other delinquent behaviors, such as taking part in a group fight, are much less common. This is interesting, but it

Table 1.2 Proportion of Adolescents Responding "Yes" to Questions About Delinquent Behaviors (Add Health Public-Use Data, Wave I; $N = 2,087$)

Questionnaire Item	Proportion Responding Yes[*]
Lied to parents	.57
Publicly loud/rowdy/unruly	.49
Damaged property	.17
Stolen something from store	.24
Stolen something worth < $50	.20
Taken part in group fight	.19

[*]Recoded from original response categories.

is possible to probe further by posing these questions: Are there distinct subgroups of adolescents within the data set that engage in particular patterns of delinquent behavior? If so, what is the distribution of adolescents across these subgroups; in other words, what are the subgroup prevalences?

To begin to address these questions, it is necessary to consider not just the individual variables, but relations among the variables. The starting point for this is the creation of a contingency table, or cross-tabulation, of the six variables. Because in this example each of the variables can take on two values ("Yes" and "No"), the contingency table has $2^6 = 64$ cells. Each cell contains a count of the number of subjects who provided a certain pattern of responses. For instance, one cell in this contingency table contains the number of adolescents who responded "No" to all six questionnaire items.

Once this contingency table is computed, what does the analyst do with it? Although examining the contingency table is a good starting point, ultimately most people would find it daunting to discern patterns in a contingency table of this size simply by inspection. LCA offers an approach to dealing with large and complex contingency tables. Using LCA it is possible to fit statistical models to these tables in order to organize and interpret the information contained there. As discussed above, each latent class model specifies some number of latent classes, which are measured by a set of observed variables. In this example, each latent class represents a group of individuals characterized by a distinct pattern of delinquent behavior. The latent classes are measured by the six delinquency items.

For now, let us set aside the issue of how the number of latent classes is arrived at and simply note that four latent classes did a satisfactory job of representing the data. (How the number of latent classes in the delinquency data was determined is discussed in Chapter 4.) The results of the LCA, presented in Table 1.3, include estimates of two sets of quantities, called *parameters*. One set of parameters contains the probability of membership in each latent class. These probabilities of latent class membership, which sum to 1 (within rounding error), are depicted graphically in Figure 1.2. (Note: Readers who try to replicate our results may wish to bear in mind

Table 1.3   Four-Latent-Class Model of Past-Year Delinquency (Add Health Public-Use Data, Wave I; $N = 2{,}087$)

	Latent Class			
Assigned label	1 Non-/Mild Delinquents	2 Verbal Antagonists	3 Shoplifters	4 General Delinquents
Probability of membership	.49	.26	.18	.06
Conditional probability of a Yes* response				
Lied to parents	.33	**.81**[†]	**.78**	**.89**
Publicly loud/rowdy/unruly	.20	**.82**	**.62**	**1.00**
Damaged property	.01	.25	.25	**.89**
Stolen something from store	.03	.02	**.92**	**.88**
Stolen something worth < $50	.00	.03	**.73**	**.88**
Taken part in group fight	.04	.31	.24	**.64**

* Recoded from original response categories.

[†] Conditional probabilities > .5 in bold to facilitate interpretation.

that their calculations will be subject to rounding error, because in this volume we report results out to only two decimal places.)

The other set of parameters represents the probabilities of each response ("Yes" or "No") to each observed variable for each latent class. Table 1.3 shows the probabilities of observing a "Yes" response on each variable, conditional on each of the four latent classes. These probabilities form the basis for interpretation and labeling of the latent classes. Ordinarily an investigator would examine these probabilities carefully before assigning labels to the latent classes. However, for purposes of this exercise, we share with the reader the labels we have assigned to the latent classes, and then show how the pattern of these probabilities is consistent with the labels. We labeled the largest latent class, which included nearly half of the subjects, Non-/Mild Delinquents. The next largest latent class, including slightly more than one-fourth of the subjects, is labeled Verbal Antagonists. The Shoplifters latent class contained about 18 percent of the subjects, and the smallest latent class, General Delinquents, contained about 6 percent of the sample.

Now let us examine the pattern of the probabilities of a "Yes" response to show how it is consistent with what one would expect, based on the labels that we have chosen for the latent classes. In Table 1.3 the larger conditional probabilities appear in bold font, to highlight the overall pattern. In addition, the pattern is shown in a graphical depiction in Figure 1.3. Latent Class 4, General Delinquents, was characterized by a high probability of responding "Yes" to all of the delinquency variables. Individuals in this latent class were likely to report having engaged in all of the delinquent behaviors listed. In contrast, those in Latent Class 1, Non-/Mild Delinquents, were likely to report not having engaged in any of the behaviors. This

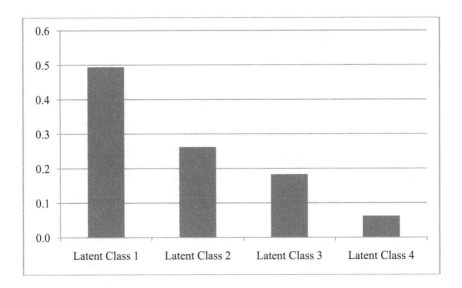

Figure 1.2   Adolescent delinquency latent class membership probabilities (Add Health public-use data, Wave I; $N = 2,087$). Note that the four probabilities sum to 1.

latent class had a somewhat higher likelihood of reporting having lied to their parents (.33) than reporting having engaged in the other behaviors, but the likelihood was still well below .5. There were two other latent classes that reflect different patterns of delinquency. Latent Class 2, Verbal Antagonists, had a high probability of reporting two types of delinquent behavior: lying to parents and public rowdiness. Those in Latent Class 3, Shoplifters, were similar to Verbal Antagonists, but in addition to lying to parents and public rowdiness, they were likely to report stealing.

It was mentioned above that categorical latent variables are characterized by qualitative differences between latent classes. This is evident in this example. The latent classes are characterized by different types of delinquent behavior. For example, shoplifting and verbal antagonism can be thought of as different domains of delinquent activity. However, there are quantitative differences as well. The four latent classes can be ordered in terms of overall involvement in delinquent behavior. Non-/Mild Delinquents had the least involvement in delinquent behavior; Verbal Antagonists were likely to report involvement in the first two behaviors only; Shoplifters were characterized by the same behaviors as Verbal Antagonists, plus the two stealing behaviors; and General Delinquents were likely to engage in all six delinquent behaviors. It is often, although by no means always, the case in LCA that there are both meaningful qualitative and quantitative differences among latent classes.

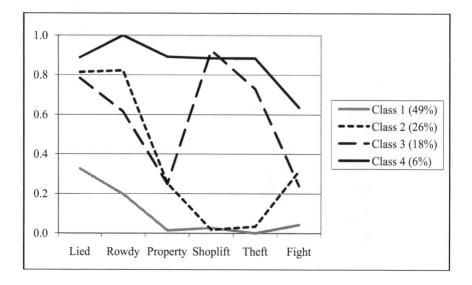

Figure 1.3   Probability of a "Yes" response to each delinquency item conditional on latent class membership (Add Health public-use data, Wave I; $N = 2,087$).

1.6   EMPIRICAL EXAMPLE OF LTA: ADOLESCENT DELINQUENCY

This book contains a substantial amount of material on latent class models for longitudinal data. Here the term *longitudinal* refers to data collected on the same individuals on at least two different occasions. When change over time is expected in a categorical or discrete phenomenon, some variation of LCA may be used to model this change. One variation of LCA for longitudinal data is LTA. In this section we present an empirical example of LTA.

The individuals in the Add Health sample who provided the data analyzed above were measured again approximately one year later. These longitudinal data make it possible to examine transitions in delinquent behavior over time. To distinguish clearly between LCA and LTA, we refer to the latent classes in LTA as *latent statuses*.

As we mentioned in Section 1.5, LCA based on a single time suggested that four latent classes represented the data well. The first step in the LTA was to examine model fit statistics to determine whether a four-latent-status solution fit the longitudinal data well, or if a model with fewer or more latent statuses would represent the data better. Even though the sample of subjects was identical to the one that provided the data analyzed in Section 1.5, the longitudinal data set was considerably different because of the addition of information from a second occasion of measurement. Additional information like this can conceivably change the optimal number of latent statuses.

(In the model fitting for this exercise we constrained the conditional probabilities for each item so that they did not change over time. This ensured that the same set of latent statuses would be identified at Times 1 and 2. (Considerations related to this choice are discussed in Chapter 7.)

We found that a five-latent-status model fit the longitudinal data best, outperforming the four-latent-status model. (As mentioned above, model selection is discussed in Chapter 4.) The five-latent-status model is shown in Table 1.4. It is interesting to note that by moving from a four-latent-class model (described above) to the five-latent-status model presented here, we essentially divided the Non-/Mild Delinquents latent class into a Nondelinquents latent status and a Liars latent status. This is discussed further below as we interpret the conditional probability of a "Yes" response to the observed delinquency variables.

The probabilities of membership in each latent status appear in the first section of Table 1.4. At Time 1 the Liars latent status was the largest, followed closely by the Nondelinquents and Verbal Antagonists latent statuses, which were about equally probable. Next was Shoplifters, which was considerably smaller. The General

Table 1.4   Five-Latent-Status Model of Past-Year Delinquency (Add Health Public-Use Data, Waves I and II; $N = 2,087$)

	Latent Status				
Assigned label	Non-delinquents	Liars	Verbal Antagonists	Shoplifters	General Delinquents
Probability of membership					
Time 1 (grades 10/11)	.24	.27	.25	.13	.10
Time 2 (grades 11/12)	.34	.28	.22	.10	.06
Conditional probability of a Yes response*					
Lied to parents	.00	**.72**[†]	**.72**	**.74**	**.89**
Publicly loud/rowdy/unruly	.15	.22	**.89**	.48	**.92**
Damaged property	.00	.05	.25	.17	**.68**
Stolen something from store	.02	.02	.04	**.92**	**.90**
Stolen something worth < $50	.00	.00	.06	**.72**	**.85**
Taken part in group fight	.03	.07	.34	.17	**.54**
Probability of transitioning to...	...Time 2 latent status				
Conditional on...					
...Time 1 latent status					
Nondelinquents	**.84**[‡]	.10	.03	.02	.01
Liars	.22	**.67**	.00	.10	.00
Verbal Antagonists	.14	.13	**.65**	.04	.04
Shoplifters	.26	.29	.11	**.34**	.00
General Delinquents	.08	.01	.36	.09	**.45**

*Recoded from original response categories.

[†]Conditional probabilities > .5 in bold to facilitate interpretation.

[‡]Diagonal transition probabilities in bold to facilitate interpretation.

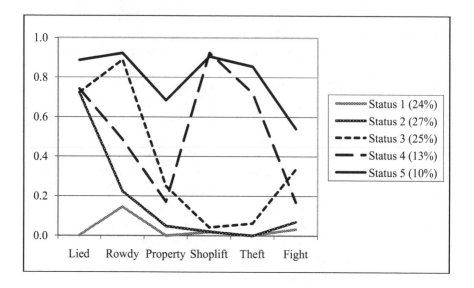

Figure 1.4 Probability of a "Yes" response to each delinquency item conditional on latent status membership (Add Health public-use data, Waves I and II; $N = 2,087$).

Delinquents latent status was the least probable. At Time 2 the overall pattern was similar, except that Nondelinquents were the most likely latent status, followed by Liars and Verbal Antagonists. The Shoplifters and General Delinquents latent statuses remained the least prevalent.

The second section of Table 1.4 contains the conditional probabilities of a "Yes" response to each delinquency item; these conditional probabilities are depicted graphically in Figure 1.4. The conditional probabilities confirm that as the labels we have chosen suggest, several of the latent statuses had counterparts in the LCA shown in Table 1.3 and Figure 1.3. The Verbal Antagonists latent status was nearly identical to the Verbal Antagonists latent class; both were characterized by a high probability of reporting lying to parents and engaging in public rowdiness. The General Delinquents latent status was also nearly identical to the General Delinquents latent class; both were characterized by a high probability of responding "Yes" to all of the delinquency items. The Shoplifters latent status in Table 1.4 was a bit different from its counterpart in Table 1.3, in that those in the Shoplifters latent status were less likely to endorse the item about public rowdiness. However, the probability of reporting shoplifting was extremely high in both cases.

Examining the conditional probabilities offers a clue about why the LTA suggested five latent statuses whereas the LCA suggested four latent classes. Table 1.3 shows that the item about lying to parents had the largest probability of endorsement (.33) in the Non-/Mild Delinquents latent class. In the LTA solution it appears that this

Non-/Mild Delinquents latent class split in two, forming Nondelinquents, who had a very low probability of endorsing any of the delinquency items, and Liars, who had a high probability (.72) of endorsing the item about lying to parents. One reason for this finer distinction may be that the additional information contained in the second time of measurement enabled more precise measurement of the latent subgroups.

The lower portion of Table 1.4 shows the transition probability matrix. This matrix expresses the probability of being in the column latent status at Time 2, conditional on being in the row latent status at Time 1. For example, the third element in the first row is .03. This means that for individuals who were in the Nondelinquents latent status at Time 1, the probability was .03 of transitioning to the Verbal Antagonists latent status. Because these transition probabilities are conditional on Time 1 latent status, each row of probabilities sums to 1. The diagonal elements of this matrix represent the probability of being in a particular latent status at Time 2 conditional on being in that same latent status at Time 1; in this example, this can be considered the probability of remaining stable in delinquency latent status over time. To assist in interpretation, the diagonal elements of the transition probability matrix appear in bold in the table.

The results in Table 1.4 suggest that Nondelinquents, Liars, and Verbal Antagonists were most likely to be members of the same latent status at Times 1 and 2, with probabilities of .84, .67, and .65, respectively. Members of the Shoplifters and General Delinquents latent statuses were much less likely to be in the same latent status at both times, with probabilities of .34 and .45, respectively. Shoplifters had a negligible (rounded to .00) probability of transitioning to the more serious latent status of General Delinquents. Instead, they were more likely to move to a latent status characterized by less severe delinquency or no delinquency. Those in the General Delinquents latent status who reduced delinquent behavior at Time 2 were most likely to transition to the Verbal Antagonists latent status.

1.7   ABOUT THIS BOOK

The purpose of this book is to provide a thorough introduction to latent class and latent transition models with categorical indicator variables. As the title of the book implies, we will emphasize the use of LCA in the social, behavioral, and health sciences. Part 1, Fundamentals, consists of the present chapter and Chapters 2–4. In this section of the book a broad introduction to latent class models is provided. Chapters 2 and 3 introduce some ideas that are essential in LCA. In Chapter 2 we introduce the standard latent class model and use two empirical examples to illustrate the basics of interpreting the results of LCA. In this chapter we also introduce the notation that is used throughout the book. Chapter 3 reviews what are desirable and undesirable characteristics of a latent class solution, and how to determine how strong the relations are between the observed variables and the latent variables. Chapter

4 covers some important technical aspects of LCA. Among these topics are how estimation is performed; identification and underidentification; how missing data are handled; and model selection.

Part 2 of the book, Advanced Latent Class Analysis, consists of Chapters 5 and 6. In Part 2 some useful variations on the standard latent class model are presented. Chapter 5 covers multiple-group LCA. As we discussed earlier in the present chapter, latent classes are unobserved subgroups in data. In addition to these unobserved subgroups, there are often meaningful *observed* subgroups in data. Examples include males and females; different cohorts, defined by age, year in school, or year of entry into a longitudinal study; and experimental conditions. It may be of scientific interest to examine whether the same latent class model can be used to represent each group, or whether there are meaningful group differences in the measurement of the latent variable. If there are group differences, multiple-group LCA can help the user to pinpoint where the differences are and to interpret them. Often, multiple-group LCA can be used to explore group differences in the prevalences of the latent classes. In Chapter 6 we show how covariates can be used to predict latent class membership by means of logistic regression. Covariates are an important interpretational tool that can help to illuminate and explain differences between latent classes. In Chapter 6 we discuss both multinomial and binomial logistic regression approaches to predicting latent class membership. We also review examining interactions between covariates and discuss how an investigator might decide whether to incorporate an exogenous variable into a latent class model as a grouping variable or a covariate.

One unique feature of this book is its emphasis on longitudinal data. This is a particular interest of ours; much of our work in the LCA area has been concerned with using this approach to express development in terms of changes in latent class membership over time. LCA for longitudinal data is covered in Part 3, which consists of Chapters 7 and 8. Chapter 7 discusses repeated-measures LCA (RMLCA), which involves using LCA to fit a repeated-measures latent class model. This model is much like a general growth mixture model (Muthén and Shedden, 1999) but does not require that the functional form of a curve be specified. The RMLCA approach can be especially helpful when there is a lot of categorical or discrete (as opposed to smooth) change. Chapter 7 also introduces LTA, which, as demonstrated in Section 1.6, is a latent class model that can be used to estimate transitions in latent class membership between successive time points. As discussed further in Chapter 7, LTA can provide a uniquely revealing look at change over time. In Chapter 8 we explore multiple-group LTA and LTA with covariates. In these models observed group membership and covariates may be used to explain and interpret transitions in latent status membership.

1.7.1 Using this book

In writing this book we assumed that the reader is familiar with basic statistical methods for the social, behavioral, and health sciences, such as regression analysis. We also assumed that the reader has a basic understanding of the laws of probability. In addition, familiarity with contingency table analysis and factor analysis is helpful. No prior exposure to LCA is necessary to read this book.

Readers who are completely unfamiliar with LCA will probably want to start at the beginning with Part I, in order to become familiar with notation, the LCA mathematical model, model selection, and other basic concepts. Readers who have had some prior exposure may wish to begin with Part II, which covers more advanced LCA models. Readers who are familiar with LCA but have not applied it to longitudinal data may be particularly interested in Part III. As much as possible, we have tried to structure the book so that it is possible to read it in sections without necessarily starting at the beginning. However, those interested primarily in Parts II and III may wish to skim Chapter 2 to become familiar with the notation used throughout the book.

At the end of the chapters there is some additional material. Most chapters include a section headed "Suggested Supplemental Readings." These suggestions are aimed at readers who may wish to delve deeper into some of topics covered. The section headed "Points to Remember" contains a brief summary of some of the key ideas that have been presented in the chapter. The section headed "What's Next" provides a transition between the current chapter and the next one.

1.8 THE EXAMPLES IN THIS BOOK

Most of the examples of LCA presented in this book are based on empirical data. We used artificial data occasionally, and clearly indicated this, when it was necessary to illustrate a point in a particularly straightforward and unambiguous way. However, we recognize that LCA of empirical data is the ultimate objective of most readers of this book. Accordingly, as much as possible we have included examples based on analyses of empirical data.

This approach resulted in examples that can be complex and, at times, perhaps a bit messy. But we know that readers will encounter complexity and messiness in their data analyses, because we encounter it in ours. Therefore, we tried to provide examples that illustrate the kinds of data analysis decisions with which our readers are likely to be faced. In many cases the reader may have made decisions slightly different from ours with respect to some aspects of the empirical analyses. We hope that even readers who disagree with our decisions will find that reading about our decision process will help make the concepts presented throughout the book more concrete.

1.8.1   Empirical data sets

In this book we rely on three large, publicly available databases for the empirical demonstrations. These data sets were selected because of their rich measurement of social, health, and behavioral constructs, their accessibility over the Internet for free or for a small charge so that interested readers may obtain the data, and their familiarity to many researchers.

1.8.1.1   Add Health

The National Longitudinal Study of Adolescent Health (Add Health; Udry, 2003) is a national longitudinal study of health and risk behavior among youth, and the influences of social factors on those behaviors. Participants were in Grades 7 through 12 at the initial assessment, which was conducted in 1994–1995. To date, three follow-up assessments have been conducted, the first approximately one year later, the second when participants were aged 18–26, and the third when there were aged 24–32. Although the public-use data set includes oversampling of a particular subpopulation, only the core participants ($N = 6,072$), a nationally representative sample, were used for examples in this book. More information about the study and how to obtain the public-use data may be found at http://www.cpc.unc.edu/addhealth.

1.8.1.2   Youth Risk Behavior Survey, 2005

The Youth Risk Behavior Survey (Centers for Disease Control and Prevention, 2005) monitors national trends in critical health risk behaviors among youth in 9th through 12th grade. School-based surveys are conducted biannually during the spring of the school year. Behaviors monitored include use of alcohol, tobacco, and other drugs, sexual behavior, and physical activity. The prevalence of obesity, asthma, and suicidal ideation are also monitored. One specific goal of the Youth Risk Behavior Survey is to examine the co-occurrence of health risk behaviors. Data from the 2005 cohort ($N = 13,917$) were analyzed in our examples. More information, as well as downloadable data and computer files, may be found at http://www.cdc.gov/HealthyYouth/yrbs/.

1.8.1.3 Monitoring the Future, 2004

Monitoring the Future: A Continuing Study of the Lifestyles and Values of Youth (Johnston et al., 2005) is an ongoing annual study of historical changes in behavior, values, and attitudes in secondary school students in the United States. Each year since 1975, a large, nationally representative sample of 12th-grade students has been surveyed. For examples in this book, we analyzed the 2004 cohort (N = 2,521). More information about the study and how to obtain the downloadable data may be found at `http://monitoringthefuture.org/`.

1.9 SOFTWARE

At this writing there are many alternatives for software to perform LCA; given the rapid pace of software development in the scientific world, by the time this book is published there will probably be a few more. For this reason, in this book we do not provide specific instructions about software or any code. All of the analyses presented in this book were performed using Proc LCA (Lanza, Collins, Lemmon, and Schafer, 2007) or Proc LTA (Lanza and Collins, 2008). Most of the analyses could also have been performed using the freeware ℓEM (`http://spitswww.uvt.nl/~vermunt/`) or the commercial software packages MPlus (`http://www.statmodel.com/`) or Latent Gold (`http://www.statisticalinnovations.com/`).

Proc LCA and Proc LTA are freeware user procs developed by us and our collaborators at The Methodology Center. They have been written to integrate seamlessly with the SAS®[1] software package and can be used to fit a wide variety of LCA and LTA models in a straightforward manner. Proc LCA and Proc LTA do not come with the standard SAS software implementation, but they can be downloaded from The Methodology Center's web site, `http://methodology.psu.edu/`. Installation of Proc LCA and Proc LTA is a simple procedure. After installation, the procedures are called in the SAS software in the same manner in which any other procedure is called.

1.10 ADDITIONAL RESOURCES: THE BOOK'S WEB SITE

We recognize that more specific information about conducing latent class and latent transition analyses can be helpful. To provide this information in a format that enables rapid updating, we have set up a special web site with supplemental information for

---

[1] SAS and all other SAS Institute Inc. product or service names are registered trademarks or trademarks of SAS Institute Inc. in the United States and other countries. ®indicates U.S. registration.

readers of this book. The web site is http://methodology.psu.edu/latentclassboo On this web site we have placed information about how to obtain the data sets used in this book, information about how to obtain the subsamples we used and how to recode the data, and Proc LCA or Proc LTA syntax. Readers are encouraged to try reproducing the analyses themselves and also to try their own variations on the analyses. Proc LCA and Proc LTA have been designed to make latent class and latent transition analyses straightforward so that they can readily be accomplished by novice and expert users alike. Readers possessing a basic familiarity with the SAS software will soon find themselves able to apply Proc LCA and Proc LTA in their own data.

1.11 SUGGESTED SUPPLEMENTAL READINGS

Von Eye and Clogg (1994) contains a variety of applications of a range of latent variable models, including factor analysis, LCA, and latent trait models. Skrondal and Rebe-Hesketh (2004) provides a more technical treatment of modern latent variable models. Hancock and Samuelsen (2008) contains a number of very readable chapters on categorical latent variable models, including Loken and Molenaar's (2008) comparison of the latent profile and factor analysis models. A variety of readings on latent class analysis, broadly defined, may be found in Hagenaars and McCutcheon (2002).

1.12 POINTS TO REMEMBER

• LCA is a latent variable model in which both the latent variable and its indicators are categorical.

• LCA is a person-oriented approach.

• LTA is a special version of LCA that is used in longitudinal data, to model transitions in latent class membership over time.

1.13 WHAT'S NEXT

This chapter has provided a conceptual, intuitive introduction to LCA and LTA. In the next chapter we provide a more specific and detailed introduction to LCA, including details of the mathematical model and statistical notation.

CHAPTER 2

THE LATENT CLASS MODEL

2.1 OVERVIEW

This chapter has two general objectives. One objective is to provide a more formal introduction to latent class analysis (LCA) than the introduction provided in Chapter 1. In addition to some simple hypothetical examples, we review two empirical examples. The empirical examples are both based on health-related data gathered on adolescents. One example is LCA of data on female pubertal development; the other is LCA of data on health risk behaviors. The examples demonstrate how two quantities are estimated in LCA. These quantities are the prevalences associated with each latent class, and the probabilities of observed responses to each item conditional on membership in each latent class.

Another objective of this chapter is to introduce the notation that will be used throughout the book and to explain the mathematical model underlying LCA. We make use of the empirical examples to illustrate how this mathematical model corresponds to conceptually interesting features of LCA.

Latent Class and Latent Transition Analysis. By Linda M. Collins and Stephanie T. Lanza
Copyright © 2010 John Wiley & Sons, Inc.

2.2 EMPIRICAL EXAMPLE: PUBERTAL DEVELOPMENT

2.2.1 An initial look at the data

We begin with an examination of data on four variables pertaining to pubertal development. The data are from the public version of the Add Health data set (Udry, 2003), Wave 1. The sample consists of 469 adolescent girls (mean age = 12.9 years) who were in the seventh grade in the United States when the data were collected in 1994–1995. Only subjects who provided a response to at least one variable were included.

Table 2.1 lists the marginal proportions corresponding to the response options for four questionnaire items. The girls were asked whether they had ever had a menstrual period; asked to compare their physical development to other girls their age; and asked to compare their current breast size and body curviness to what it was in grade school. The question about menstruation was a yes/no question, whereas the remaining questions were recoded so that each had three response alternatives. As is noted in Table 2.1, like most empirical data sets the Add Health data set has some unplanned missing data that occurred because, for example, some subjects may have chosen not to respond to some questions.

A few general trends are evident in the marginals shown in Table 2.1. Most of the girls (70%) had had a menstrual period. Forty percent felt that compared to girls

Table 2.1   Marginal Response Proportions for Female Pubertal Development Variables (Add Health Public-Use Data, Wave I; $N = 469$)

Indicator of Pubertal Development	Response Proportion
Ever had menstrual period	
No	.30
Yes	.70
How advanced is your physical development compared to other girls your age?	
Look younger than most/some	.24
Look about average	.40
Look older than most/some	.36
Compared to grade school, my breasts are...	
Same size/a little bigger	.39
Somewhat bigger	.36
A lot/a whole lot bigger	.25
Compared to grade school, my body is...	
As curvy/a little more curvy	.41
Somewhat more curvy	.33
A lot/a whole lot more curvy	.26

Note. Proportions are based on N responding to each question. The amount of missing data varied slightly across questions.

their age, their physical development was about average. When they compared their breasts and body to what they looked like in grade school, most of the girls noted relatively little change. Despite these general trends, there is substantial variability. Thirty percent of the girls had not had a menstrual period, and each of the response categories in the remaining three questionnaire items was selected by no fewer than 24 percent of the subjects.

Given that there is so much variability evident in Table 2.1, a next question might be whether there are any patterns of responses that stand out as occurring more frequently than the others. To see this, it is necessary to cross-tabulate the four variables and create a contingency table. In the case of the data set with which we are presently working, this is a fairly large contingency table. Because there is one variable with two response alternatives and three variables with three response alternatives, the size of the contingency table is $2 \times 3 \times 3 \times 3 = 54$ cells. Each of the cells of the contingency table corresponds to a combination of responses to the four questionnaire items, or a *response pattern*, and contains the frequency (or, equivalently, proportion) of subjects who provided that response pattern.

Table 2.2 contains the response patterns and response pattern frequencies for the pubertal development data. In this book response patterns appear in parentheses with the individual item responses separated by commas. To save space, Table 2.2 uses numbers to designate each response alternative. For example, Table 2.2 shows that 36 subjects provided response pattern (1,1,1,1), which represents a response pattern made up of the first response alternative to each of the four questionnaire items. In the pubertal development example, this is equivalent to (No, Look younger than most/some, Same size/a little bigger, As curvy/a little more curvy).

The upper portion of Table 2.2 lists the complete response patterns that were observed in the data set. In a complete response pattern there are data on all of the variables. We noted above that in this example, the contingency table has 54 cells. In Table 2.2 51 complete response patterns are listed, each of which corresponds to a cell of the contingency table. The remaining three cells of the contingency table have an observed frequency of zero, in other words, are empty; no subjects in this sample provided those sets of responses. The response patterns corresponding to these empty cells are not listed in the table.

The lower portion of Table 2.2 lists the 14 incomplete response patterns that were observed in the data set. In an incomplete response pattern, data are missing on one or more of the variables. In the incomplete response patterns listed in Table 2.2, the missing responses are indicated by a period. Incomplete response patterns may occur because a subject did not respond to one or more questionnaire items, some responses were illegible, there was a data transcription error, and so on. Most empirical data sets in the social, behavioral, and health sciences include both complete and incomplete response patterns.

Table 2.2   Response Patterns and Frequencies for Add Health Pubertal Development Data (Add Health Public-Use Data, Wave I; $N = 469$)

Response Pattern	Frequency	Response Pattern	Frequency
	Complete response patterns		
(1, 1, 1, 1)	36	(2, 1, 1, 2)	4
(1, 1, 1, 2)	3	(2, 1, 2, 1)	3
(1, 1, 1, 3)	3	(2, 1, 2, 2)	9
(1, 1, 2, 1)	7	(2, 1, 2, 3)	2
(1, 1, 2, 2)	8	(2, 1, 3, 1)	1
(1, 1, 2, 3)	1	(2, 1, 3, 2)	1
(1, 1, 3, 2)	3	(2, 1, 3, 3)	4
(1, 1, 3, 3)	1	(2, 2, 1, 1)	26
(1, 2, 1, 1)	17	(2, 2, 1, 2)	13
(1, 2, 1, 2)	5	(2, 2, 1, 3)	4
(1, 2, 1, 3)	1	(2, 2, 2, 1)	24
(1, 2, 2, 1)	8	(2, 2, 2, 2)	29
(1, 2, 2, 2)	14	(2, 2, 2, 3)	9
(1, 2, 2, 3)	2	(2, 2, 3, 1)	2
(1, 2, 3, 2)	4	(2, 2, 3, 2)	6
(1, 2, 3, 3)	2	(2, 2, 3, 3)	13
(1, 3, 1, 1)	3	(2, 3, 1, 1)	19
(1, 3, 1, 2)	1	(2, 3, 1, 2)	7
(1, 3, 1, 3)	2	(2, 3, 1, 3)	8
(1, 3, 2, 1)	2	(2, 3, 2, 1)	7
(1, 3, 2, 2)	4	(2, 3, 2, 2)	22
(1, 3, 2, 3)	2	(2, 3, 2, 3)	11
(1, 3, 3, 1)	1	(2, 3, 3, 1)	4
(1, 3, 3, 2)	1	(2, 3, 3, 2)	18
(1, 3, 3, 3)	4	(2, 3, 3, 3)	46
(2, 1, 1, 1)	23		
	Incomplete response patterns		
(. , . , . , 2)	1	(2, . , 1, 1)	1
(. , . , 1, 1)	1	(2, 2, . , .)	2
(. , 2, 2, 1)	2	(2, 2, 2, . .)	1
(. , 3, 1, 1)	1	(2, 2, 3, .)	1
(. , 3, 3, 3)	1	(2, 3, . . , 1)	1
(1, 1, . , .)	2	(2, 3, . . , 3)	1
(1, 2, . , .)	2	(2, 3, 3, . .)	2

Note. A period (.) represents missing data.

When the number of cells in the contingency table is computed, incomplete response patterns are not counted; in other words, the size of the contingency table is based on the number of possible complete response patterns only. This is important to note because as will be seen later in the book, computation of the number of degrees of freedom, which is necessary for hypothesis testing, starts with the number of cells in the contingency table. Although the incomplete response patterns are not involved in computation of either the size of the contingency table or the number of degrees of freedom, they may be included in a data set that is to be fit with a latent class model; in fact, the incomplete response patterns listed in Table 2.2 were included in the analyses reported here. (Treatment of missing data in LCA is discussed in Chapter 4.)

Examining which of the complete response patterns occur most frequently can be a helpful way to gain some initial familiarity with empirical data, although if there are a lot of missing data, or the missing data diverge strongly from the "missing completely at random" (MCAR) assumption (Schafer, 1997; Schafer and Graham, 2002), this can sometimes be misleading. The four response patterns that occur most frequently in Table 2.2 are: menstruation and advanced physical development, including breast and body development (2, 3, 3, 3), with an observed frequency of 46; little pubertal development (1, 1, 1, 1), with an observed frequency of 36; menstruation, average physical development, and some breast and body development (2, 2, 2, 2), with an observed frequency of 29; and menstruation and average physical development, but no breast and body development (2, 2, 1, 1), with an observed frequency of 26.

2.2.2   Why conduct LCA on the pubertal development data?

Why conduct LCA on the pubertal development data? One important motivation is the need for a way to represent the complex array of data in Table 2.2 in a format that is more parsimonious and easier to comprehend, and at the same time reveals important scientific information contained in the data. Although it is interesting to note the trends in the data discussed in the preceding section, together the four most frequently occurring response patterns represent only 137 subjects, or less than 30 percent of the sample. The remaining 70 percent of the sample have contributed other response patterns, ranging in frequency from 24 to 1. Thus the most frequently occurring response patterns are an initial hint about what scientific information the data contain, but much information is provided elsewhere in the data as well. LCA will enable us to fit a model that represents the entire data set, including both common and uncommon response patterns, as well as complete and incomplete response patterns.

As discussed in Chapter 1, one idea behind LCA is that two distinct quantities determine an individual's observed responses. One quantity is the individual's true latent class. The other quantity is error. Because of this error, sometimes an individual's observed responses will not reflect his or her true latent class perfectly. With so many response patterns and the influence of error on the observed data, it is difficult for the unaided human mind to identify the scientifically important trends or patterns in the data. The purpose of LCA is to help the investigator to discern any meaningful, scientifically interesting latent classes against the noisy background of error. This is similar in spirit to the goal of any latent variable approach.

We will return to the concept of error again in this chapter and in the following chapter. For now, keep in mind that **the overall objective of performing a latent class analysis on a set of variables is to arrive at an array of latent classes that represents the response patterns in the data, and to provide a sense of the prevalence of each latent class and the amount of error associated with each variable in measuring these latent classes.**

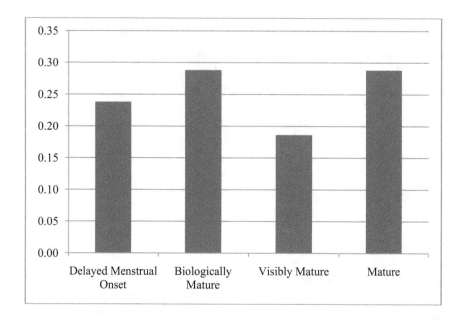

Figure 2.1   Latent class prevalences in pubertal development example (Add Health public-use data, Wave I; $N = 469$).

2.2.3   Latent classes in the pubertal development data

LCA performed on the pubertal development data indicated that four latent classes represented the data adequately. (How this number of latent classes was selected and how in general a user decides on the number of latent classes are discussed in Chapter 4.) Figure 2.1 is a bar chart illustrating the estimated prevalence of each of the latent classes, in other words, the estimated number of individuals in each latent class. As Figure 2.1 shows, the two largest latent classes are Biologically Mature and Mature, each containing approximately 29 percent of the sample. About 24 percent of the girls are in the Delayed Menstrual Onset latent class. The smallest latent class, Visibly Mature, contains about 19 percent of the sample.

Note that the latent class prevalences sum to 100 percent (within rounding). This is because the latent classes are mutually exclusive and exhaustive. In other words, the underlying idea is that everyone belongs to one and only one latent class. No one belongs to more than one latent class, so the sum cannot exceed 100 percent, and everyone is classified, so the sum of the prevalences cannot be less than 100 percent. However, it is important to note that according to the latent class model, this classification is probabilistic. In other words, the model does not "know" each individual's latent class membership. Instead, corresponding to each individual there is a vector of probabilities of membership in each latent class. This is discussed further in Chapter 3.

2.3 THE ROLE OF ITEM-RESPONSE PROBABILITIES IN INTERPRETING LATENT CLASSES

Where did the labels Delayed Menstrual Onset, Biologically Mature, Visibly Mature, and Mature come from? When a latent class analysis is performed, labels do not appear on the computer output! Just as in factor analysis where the investigator must assign labels to the factors, in LCA it is the responsibility of the investigator to assign names to the latent classes. As discussed in Chapter 1, interpretation of the latent classes is based on the *item-response probabilities*. Before examining the item-response probabilities for the pubertal development example, let us review a brief hypothetical example of assigning labels to latent classes based on item-response probabilities.

2.3.1 A hypothetical example

Suppose that a brief test consisting of three practical tasks is administered to a sample of participants. Performance on each task is rated as fail or pass. A model with two latent classes is fit, producing the estimates of item-response probabilities that appear in Table 2.3. Each item-response probability is the probability of a particular response to a particular task, or item, conditional on membership in a particular latent class. For example, the first item-response probability in the first row means that individuals in Latent Class 1 have a probability of .1 of failing Task 1. Similarly, individuals in Latent Class 2 have a probability of .9 of failing Task 1.

The investigator must now interpret the meaning of the latent classes based on these estimated item-response probabilities. The estimates indicate that those in Latent Class 1 have a high probability of passing all these tasks successfully, whereas those in Latent Class 2 have a low probability of passing the tasks successfully. Based on this overall pattern of item-response probabilities, Latent Class 1 might be labeled Masters and Latent Class 2 might be labeled Nonmasters.

Suppose instead that a three-latent-class model was fit, producing the estimates of item-response probabilities in Table 2.4. This pattern of item-response probabilities suggests that in addition to a Masters (Latent Class 1) and a Nonmasters (Latent Class 3) latent class, there is another class, made up of individuals who have a high probability of passing Tasks 1 and 2 but have a low probability of passing Task 3. Latent Class 2 could perhaps be labeled Intermediates. Such a model would suggest that Task 3 is more difficult than Tasks 1 and 2.

Table 2.3   Item-Response Probabilities for a Hypothetical Two-Latent-Class Model

Practical Task	Latent Class 1	Latent Class 2
Task 1		
Fail	.1	.9
Pass	.9	.1
Task 2		
Fail	.1	.9
Pass	.9	.1
Task 3		
Fail	.1	.9
Pass	.9	.1

Table 2.4   Item-Response Probabilities for a Hypothetical Three-Latent-Class Model

Practical Task	Latent Class 1	Latent Class 2	Latent Class 3
Task 1			
Fail	.1	.1	.9
Pass	.9	.9	.1
Task 2			
Fail	.1	.1	.9
Pass	.9	.9	.1
Task 3			
Fail	.1	.9	.9
Pass	.9	.1	.1

2.3.2   Interpreting the item-response probabilities to label the latent classes in the pubertal development example

The item-response probabilities for the pubertal development example, along with the latent class prevalences, appear in Table 2.5. For example, for a girl in the Delayed Menstrual Onset latent class, the probability was .64 of responding "No" to the question about ever having had a menstrual period. As demonstrated in the preceding section, the overall pattern of item-response probabilities for a particular latent class informs the choice of label for that latent class.

Figures 2.2 through 2.5 contain the same item-response probabilities that appear in Table 2.5, but in bar chart form. Each of the four figures corresponds to one of the latent classes. As we review each figure, try to imagine that the latent class labels have not yet been assigned and must now be determined based on the item-response probabilities.

The girls in the latent class represented in Figure 2.2 were likely to report that they had never had a menstrual period, that they looked younger than other girls their age, and that compared to grade school their breasts were the same size or a little bigger

Table 2.5   Four-Latent-Class Model of Female Pubertal Development in Seventh Grade (Add Health Public-Use Data, Wave I; $N = 469$)

	Latent Class			
	Delayed Menstrual Onset	Biolog- ically Mature	Visibly Mature	Mature
Latent class prevalences	.24	.29	.19	.29
Item-response probabilities				
Ever had menstrual period				
No	**.64**[*]	.16	.41	.09
Yes	.36	**.84**	**.59**	**.91**
How advanced is your physical development compared to that of other girls your age?				
Look younger than most/some	**.73**	.00	.27	.05
Look about average	.24	**.65**	**.52**	.20
Look older than most/some	.02	.35	.21	**.75**
Compared to grade school, my breasts are...				
Same size/a little bigger	**.85**	**.53**	.03	.10
Somewhat bigger	.15	.47	**.80**	.15
A lot/a whole lot bigger	.00	.00	.17	**.75**
Compared to grade school, my body is...				
As curvy/a little more curvy	**.87**	**.63**	.02	.08
Somewhat more curvy	.09	.30	**.86**	.22
A lot/a whole lot more curvy	.04	.07	.12	**.70**

[*] Item-response probabilities > .5 in bold to facilitate interpretation.

and their body was as curvy or a little more curvy. Conversely, they were unlikely to report having had a menstrual period, that their physical development was more advanced than or even about average compared to other girls their age, or that there had been much breast or body development since grade school. This overall pattern suggests that this latent class could be labeled Delayed Menstrual Onset.

Figure 2.3 shows a different pattern. The girls in this latent class were likely to report that they had had a menstrual period and that their physical development was about average for girls their age. Compared to the girls in the Delayed Menstrual Onset latent class, they were somewhat more likely to report breast and body development since grade school, although they still were most likely to report that there had been no or a little breast and body development. These girls were characterized by biological maturity in terms of menstruation and perhaps other aspects of physical development, but were not yet visibly very developed. Accordingly, we label this latent class Biologically Mature.

Figure 2.4 shows that in the third latent class, girls had a probability of about .59 of reporting having had a menstrual period. This means that although they were more likely than not to report having menstruated, they were less likely to report this than were the girls in the Biologically Mature latent class. For the question

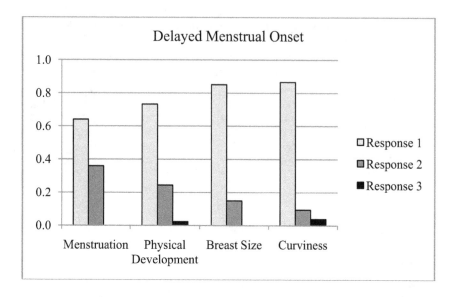

Figure 2.2   Item-response probabilities for measurement of Delayed Pubertal Onset latent class (Add Health public-use data, Wave I; $N = 469$). Response category labels appear in Table 2.5.

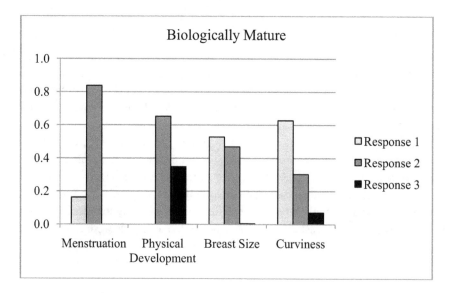

Figure 2.3   Item-response probabilities for measurement of Biologically Mature latent class (Add Health public-use data, Wave I; $N = 469$). Response category labels appear in Table 2.5.

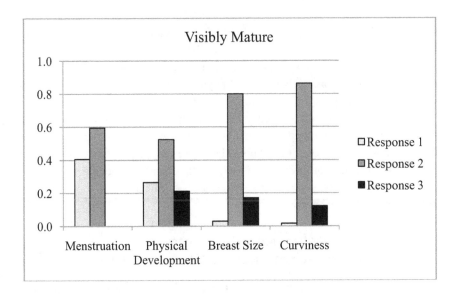

Figure 2.4   Item-response probabilities for measurement of Visibly Mature latent class (Add Health public-use data, Wave I; $N = 469$). Response category labels appear in Table 2.5.

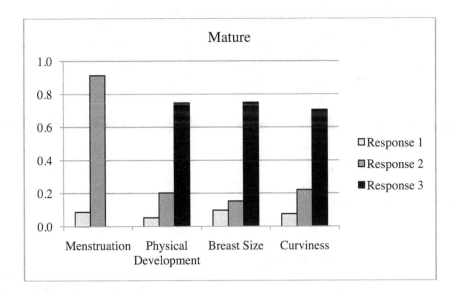

Figure 2.5   Item-response probabilities for measurement of Mature latent class (Add Health public-use data, Wave I; $N = 469$). Response category labels appear in Table 2.5.

asking about physical development compared to others their age, girls in this latent class were most likely to report that their physical development was about average. However, compared to the Biologically Mature latent class, they were more likely to report visible breast and body development since grade school. These girls were characterized primarily by visible signs of physical development. Thus a possible label for this latent class is Visibly Mature.

The girls in the latent class profiled in Figure 2.5 were very likely to report having had a menstrual period, looking older than girls their age, and having a lot of breast and body development since grade school. They were very unlikely to report looking younger than other girls or having no or only a little breast and body development. This latent class was characterized by a high level of pubertal development, so it seems reasonable to label it Mature.

2.3.3   Qualitative and quantitative differences among the pubertal development latent classes

In Chapter 1 it was noted that there are often qualitative differences among latent classes. One way to think of this is that the subgroups revealed in LCA may not be ordered along a single quantitative dimension. In the present example the classes can be roughly ordered to reflect increasing pubertal development, but that is not the entire story. The Mature latent class was ahead of the other three, and the Biologically Mature and Visibly Mature girls were ahead of the girls in the Delayed Menstrual Onset latent class. However, the Biologically Mature and Visibly Mature latent classes cannot readily be ordered in this way. These two latent classes of girls appeared to be developing in different ways. The Biologically Mature girls were likely to have started menstruating but unlikely to have undergone changes in their breasts and body. In contrast, the Visibly Mature girls were less likely to have started menstruating but very likely to have undergone changes in their breasts and body. Thus there is a rough quantitative dimension of overall pubertal development underlying some of the latent classes, but there are also more complex, qualitative differences among them.

2.4   EMPIRICAL EXAMPLE: HEALTH RISK BEHAVIORS

2.4.1   An initial look at the data

To gain experience in using the item-response probabilities to interpret the meaning of latent classes, let us examine another empirical example. This one is based on data about 12 health risk behaviors gathered from 13,840 U.S. high school students who were in the 2005 cohort of the Youth Risk Behavior Survey (Centers for Disease Control and Prevention, 2004). The students, who were in Grades 9 through 12, were

Table 2.6   Proportion of Students Reporting Each Health Risk Behavior (Youth Risk Behavior Survey, 2005; $N = 13,840$)

Health Risk Behavior	Proportion Responding Yes
Smoked first cigarette before age 13	.15
Smoked daily for 30 days	.12
Has driven when drinking	.11
Had first drink before age 13	.26
≥ 5 drinks in a row in past 30 days	.25
Tried marijuana before age 13	.09
Used cocaine in life	.08
Sniffed glue in life	.12
Used meth in life	.06
Used ecstasy in life	.06
Had sex before age 13	.07
Had sex with four or more people	.17

Note. Proportions are based on N responding to each question. The amount of missing data varied across questions.

asked to indicate whether they had engaged in each behavior. The 12 health risk behaviors and the proportions of students who responded "Yes," indicating that they had engaged in the behavior, are shown in Table 2.6.

Several general trends are evident in Table 2.6. Alcohol use was the most frequently endorsed health risk behavior, with 26 percent of the sample reporting having had a first drink before age 13 and 25 percent reporting five or more drinks in a row in the past 30 days. The next most frequently endorsed item was having had sex with four or more people (17%). Fifteen percent of the youth reported smoking a first cigarette before age 13. Daily smoking for 30 days and having ever sniffed glue were each endorsed by 12 percent of the sample. The questions endorsed by the smallest proportion of the sample concerned ever using methamphetamines or ecstasy.

Table 2.6 shows that the different health risk behaviors were endorsed by different proportions of individuals. However, it does not show whether there are subgroups of youth who are characterized by engagement in different types of risky behavior. Perhaps LCA can be helpful in illuminating this complex array of data.

2.4.2 LCA of the health risk behavior data

LCA of the health risk behavior data suggested that five latent classes represented the data well. (More about selection of this model appears in Chapter 4.) Before we reveal our labels for the latent classes, we would like the reader to engage in the exercise of examining the patterns of item-response probabilities and assigning labels to the latent classes. To assist in interpretation of the latent classes, we have created bar charts of the estimated item-response probabilities corresponding to the "Yes" response for each of the five latent classes. Because there are so many variables in this empirical example, we have created three bar charts. Figure 2.6 shows the item-response probabilities for the questions about alcohol and tobacco use; Figure

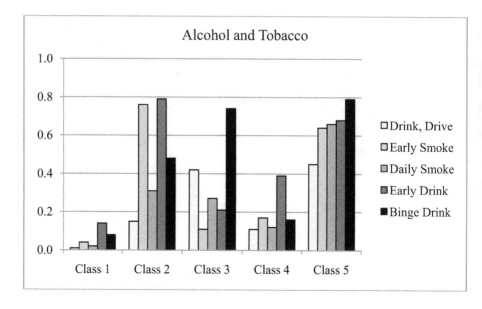

Figure 2.6   Probability of endorsing alcohol and tobacco use items conditional on latent class membership (Youth Risk Behavior Survey, 2005; $N = 13,840$).

2.7 shows the item-response probabilities for the questions about other drug use; and Figure 2.8 shows the item-response probabilities for the questions concerning sexual behavior.

As the figures show, individuals in the first latent class had a low probability of indicating that they had engaged in any of the health risk behaviors. This suggests that a label of Low Risk is appropriate. Individuals in the second latent class differed from the first in their alcohol and tobacco use: They were likely to have smoked a cigarette and to have had a drink before age 13. Their pattern of responses on the remaining variables was similar to that of the first latent class. The second latent class can be labeled Early Experimenters. Those in the third latent class had a high probability of responding that they had had five or more drinks in a row at least once in the last 30 days, and were somewhat more likely than those in latent classes 1, 2, and 4 to have driven when drinking. Their other drug use and sexual behavior was similar to that of the first two latent classes. An appropriate label for this latent class is Binge Drinkers. The risky behaviors of individuals in the fourth latent class mainly concerned sexual activity; these individuals had a relatively high probability of having had sex before age 13 and of having had four or more sex partners in their life. They were also somewhat more likely than those in latent classes 1 and 3, although less likely than those in latent classes 2 and 5, to have had a first drink before age 13. This latent class can be labeled Sexual Risk-Takers. Those in latent class 5

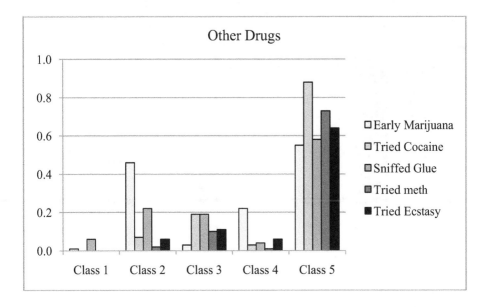

Figure 2.7   Probability of endorsing other drug use items conditional on latent class membership (Youth Risk Behavior Survey, 2005; $N = 13,840$).

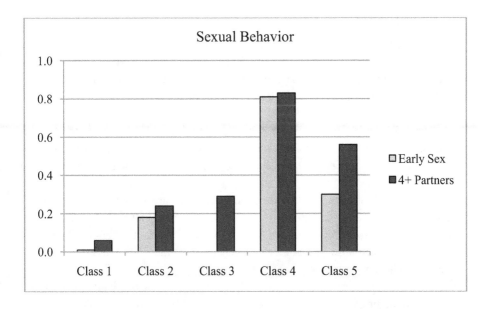

Figure 2.8   Probability of endorsing sexual behavior items conditional on latent class membership (Youth Risk Behavior Survey, 2005; $N = 13,840$).

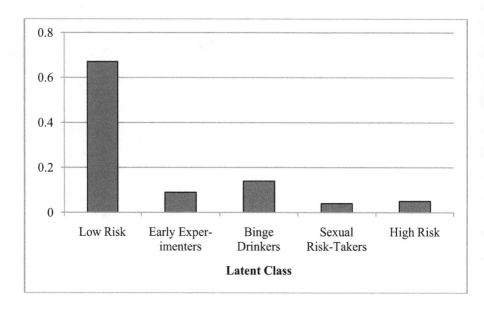

Figure 2.9   Prevalence of health risk behavior latent classes (Youth Risk Behavior Survey, 2005; $N = 13{,}840$).

were likely to have engaged in most of the risk behaviors. Behaviors they were least likely to report include sex before age 13 and driving while drinking, although they were more likely than individuals in any of the other latent classes to have engaged in the latter behavior. This latent class will be labeled High Risk.

Figure 2.9 shows the latent class prevalences. Recall that these prevalences sum to 1 because the latent classes are mutually exclusive and exhaustive (see Section 2.2.3). The Low Risk latent class is the most prevalent. The Binge Drinkers latent class is the next most prevalent, followed by Early Experimenters, High Risk, and Sexual Risk-Takers.

Table 2.7 shows the latent class prevalences that are illustrated in Figure 2.9 and the item-response probabilities that are illustrated in Figures 2.6, 2.7, and 2.8. Item-response probabilities greater than .5 are in bold font. Table 2.7 contains only the item-response probabilities associated with the "Yes" response for each variable. The item-response probabilities corresponding to the "No" response are the complements of these (i.e., as Equation 2.2 indicates, they can be found by subtracting the corresponding item-response probabilities in Table 2.7 from 1). When there are only two response alternatives for all variables in LCA, we find that an abbreviated table of item-response probabilities consisting only of those associated with one of the response alternatives (most often a "Yes" or positive response) is usually sufficient

Table 2.7   Five-Latent-Class Model of Health Risk Behaviors (Youth Risk Behavior Surveillance System Data; $N = 13,840$)

	Latent Class				
	Low Risk	Early Experimenters	Binge Drinkers	Sexual Risk-takers	High Risk
Latent class prevalences	.67	.09	.14	.04	.05
Item-response probabilities	Probability of a Yes response				
Smoked first cigarette before age 13	.04	**.76**[*]	.11	.17	**.64**
Smoked daily for 30 days	.02	.31	.27	.12	**.66**
Has driven when drinking	.01	.15	.42	.11	.45
Had first drink before age 13	.14	**.79**	.21	.39	**.68**
$\geq$ 5 drinks in a row in past 30 days	.08	.48	**.74**	.16	**.79**
Tried marijuana before age 13	.01	.46	.03	.22	**.55**
Used cocaine in life	.00	.07	.19	.03	**.88**
Sniffed glue in life	.06	.22	.19	.04	**.58**
Used meth in life	.00	.02	.10	.01	**.73**
Used ecstasy in life	.00	.06	.11	.06	**.64**
Had sex before age 13	.01	.18	.00	**.81**	.30
Had sex with four or more people	.06	.24	.29	**.83**	**.56**

[*] Item-response probabilities > .5 in bold to facilitate interpretation.

Note. The probability of a "No" response can be calculated by subtracting the item-response probabilities shown above from 1.

for purposes of interpretation. When there are more than two response alternatives for any variable, a complete table like Table 2.5 is more helpful.

2.5   LCA: MODEL AND NOTATION

In this section we introduce the latent class model in a more formal and general way. As demonstrated above, the starting point for conducting a latent class analysis on empirical data is a contingency table formed by cross-tabulating all the observed variables to be involved in the analysis, in much the same way that a covariance or correlation matrix is the starting point for factor analysis. A latent class model is made up of estimated latent class prevalences and item-response probabilities that can be used to obtain expected cell proportions for this contingency table. If the model fits the data well, the expected cell proportions closely match the observed cell proportions. To simplify the exposition, the presentation of the LCA model assumes that no missing data on the observed indicator variables. (As mentioned above, evaluating the fit of a model and dealing with missing data are discussed in Chapter 4.)

Suppose that there are $j = 1, ..., J$ observed variables, and observed variable j has $r_j = 1, ..., R_j$ response categories. The contingency table formed by cross-tabulating the J variables has $W = \prod_{j=1}^{J} R_j$ cells. In the pubertal development example discussed above, the observed variables are the questionnaire items, so $J = 4$. For the first questionnaire item (Ever had menstrual period) there are two response alternatives, "No" and "Yes," so $R_1 = 2$; for the second questionnaire item (How advanced is your physical development compared to that of other girls your age?) there are three response alternatives, so $R_2 = 3$; and so on. As shown above, in the pubertal development example, $W = 2 \times 3 \times 3 \times 3 = 54$. In the health risk behaviors example, there are 12 questionnaire items, each with two response alternatives. Therefore, in that example, $W = 2^{12} = 4,096$.

Corresponding to each of the W cells in the contingency table is a complete response pattern, which is a vector of responses to the J variables, represented by $\mathbf{y} = (r_1, ..., r_J)$. For instance, in the pubertal development example a response pattern of $(2, 1, 2, 3)$ represents responses of "Yes" to the first questionnaire item, "Look younger than most/some" to the second, "Somewhat bigger" to the third, and "A lot/a whole lot more curvy" to the last. Let $\mathbf{Y}$ refer to the array of response patterns. $\mathbf{Y}$ has W rows and J columns. Each response pattern $\mathbf{y}$ is associated with probability $P(\mathbf{Y} = \mathbf{y})$, and $\sum P(\mathbf{Y} = \mathbf{y}) = 1$.

In the more formal model the latent class prevalences and item-response probabilities are referred to as *parameters* and are represented by Greek letters. The latent class prevalences are referred to as γ's (gamma's) and the item-response probabilities are referred to as ρ's (rho's). (A reminder: Here the Greek letter ρ is used to represent something different from a correlation.) Because there are several γ's and several ρ's in latent class models, we sometimes refer to sets of γ and ρ parameters. In understanding the mathematical model for LCA, it helps to note that the γ and ρ parameters are probabilities and therefore obey all the usual laws of probability. How the γ and ρ parameters can be estimated based on empirical data is discussed in Chapter 4.

Let L represent the categorical latent variable with $c = 1, ..., C$ latent classes. γ_c represents the prevalence of latent class c, or in other words, the probability of membership in latent class c of latent variable L. In the pubertal development example there are four γ's, one corresponding to each latent class. For instance, $\gamma_2 = .29$ represents the probability of membership in Latent Class 2, Biologically Mature. As mentioned above, the latent classes are mutually exclusive and exhaustive; in other words, each individual is a member of one and only one latent class. Therefore,

$$\sum_{c=1}^{C} \gamma_c = 1. \tag{2.1}$$

The item-response probability $\rho_{j,r_j|c}$ represents the probability of response r_j to observed variable j, conditional on membership in latent class c. For example, $\rho_{1,2|2} = .84$ represents the probability that if a girl is in Latent Class 2, Biologically Mature, she will provide response "Yes" to the first questionnaire item (Ever had menstrual period). The set of ρ parameters expresses the relation between each observed indicator variable and each latent class. As a set, together these parameters express how well individuals can be classified into latent classes given the set of observed variables.

In addition to the term item-response probability, the ρ parameters are sometimes referred to as measurement parameters. This is because, as was demonstrated above, the ρ parameters are key to the measurement of the latent variable and the interpretation of latent classes. We pay considerably more attention to the ρ parameters in Chapter 3.

Because each individual provides one and only one response alternative to variable j, the vector of item-response probabilities for a particular variable conditional on a particular latent class always sums to 1. Thus

$$\sum_{r_j=1}^{R_j} \rho_{j,r_j|c} = 1 \tag{2.2}$$

for all j. For example, the probabilities associated with the various response alternatives for the fourth question conditional on membership in the third latent class, Visibly Mature, are: $\rho_{4,1|3}$ (i.e., for the response "As curvy/a little more curvy") = .02; $\rho_{4,2|3}$ (i.e., for the response "Somewhat more curvy") = .86; and $\rho_{4,3|3}$ (i.e., for the response "A lot/a whole lot more curvy") = .12. Then $.02 + .86 + .12 = 1$. (When based on any of the tables in this book, which show the parameter estimates to only two decimal places, this sum may not equal 1 exactly, due to rounding error. The original estimates sum to 1 to at least 15 decimal places.)

2.5.1 Fundamental expressions

Now we examine several mathematical expressions that are fundamental to latent class models. Let y_j represent element j of a response pattern $\mathbf{y}$. Let us establish an indicator function $I(y_j = r_j)$ that equals 1 when the response to variable $j = r_j$, and equals 0 otherwise. (As will be evident below, this function is merely a device for picking out the appropriate ρ parameters to multiply together.) Then Equation 2.3 expresses how the probability of observing a particular vector of responses is a function of the probabilities of membership in each latent class (the γ's) and the probabilities of observing each response conditional on latent class membership (the ρ's):

$$P(\mathbf{Y} = \mathbf{y}) = \sum_{c=1}^{C} \gamma_c \prod_{j=1}^{J} \prod_{r_j=1}^{R_j} \rho_{j,r_j|c}^{I(y_j=r_j)}. \tag{2.3}$$

Taking Equation 2.3 apart may help in understanding it. We will refer to a small, simple hypothetical example. Suppose that a group of children has been given two math problems to solve, each of which can be passed or failed. Suppose further that there are two latent classes: Below Grade Level and At Grade Level. Those in the Below Grade Level latent class do not have the skills to perform the math problems and thus are likely to fail both, whereas those in the At Grade Level latent class have acquired the necessary skills and thus are expected to pass both math problems. Table 2.8 shows latent class prevalences and item-response probabilities for this example.

Table 2.8   Hypothetical Example with Two Latent Classes and Two Observed Variables

	Below Grade Level	At Grade Level
Latent class prevalences	.3	.7
Item-response probabilities		
Math problem 1		
Fail	.8	.1
Pass	.2	.9
Math problem 2		
Fail	.8	.1
Pass	.2	.9

With two pass/fail math problems there are four possible response patterns: (Fail, Fail), (Fail, Pass), (Pass, Fail), and (Pass, Pass). Suppose that we wish to use Equation 2.3 to compute $P(\mathbf{Y} = (\text{Pass, Pass}))$ based on the parameters in Table 2.8. To look at Equation 2.3 more closely, let us work from right to left. The second factor, $\prod_{j=1}^{J} \prod_{r_j=1}^{R_j} \rho_{j,r_j|c}^{I(y_j=r_j)}$, represents the probability of a particular observed response pattern y conditional on membership in latent class c:

$$P(\mathbf{Y} = \mathbf{y}|L = c) = \prod_{j=1}^{J} \prod_{r_j=1}^{R_j} \rho_{j,r_j|c}^{I(y_j=r_j)}. \tag{2.4}$$

For example, $P(\mathbf{Y} = (\text{Pass, Pass})|\text{At Grade Level})$ represents the probability of response pattern (Pass, Pass) conditional on membership in the At Grade Level latent class. Equation 2.4 can also be considered the probability of response r_1 to observed variable 1 *and* response r_2 to observed variable 2 *and...and* response r_J to observed variable J, all conditional on membership in latent class c. In terms of the hypothetical example, this would be the probability of passing Math Problem 1 *and* passing Math Problem 2, conditional on membership in the At Grade Level latent class. To obtain this probability it is necessary to multiply the probabilities of all the individual responses. Thus, referring to Table 2.8, $P(\mathbf{Y} = (\text{Pass, Pass})|\text{At Grade Level}) = \rho_{1,\text{Pass}|\text{At Grade Level}} \times \rho_{2,\text{Pass}|\text{At Grade Level}} = .9 \times .9 = .81$. In other words, for those who are in the At Grade Level latent class, the probability of passing both math problems is .81.

Of course, as mentioned above, it is expected that the probability of passing both math problems is high for those who are in the At Grade Level latent class. However, there is also a nonzero probability of passing both math problems for those who are in the Below Grade Level latent class. Thus, similarly, $P(\mathbf{Y} = (\text{Pass, Pass})|\text{Below Grade Level}) = \rho_{1,\text{Pass}|\text{Below Grade Level}} \times \rho_{2,\text{Pass}|\text{Below Grade Level}} = .2 \times .2 = .04$. As one might expect, this is a small probability, but it cannot be ignored. In general, to compute the overall probability of a particular response pattern, it is necessary to consider the conditional probability of the response pattern for all of the latent classes.

After computing $P(\mathbf{Y} = \mathbf{y}|L = c)$ for each latent class c, the next step toward obtaining $P(\mathbf{Y} = \mathbf{y})$ is to compute another set of intermediate quantities. These are the unconditional *joint* probabilities of latent class c *and* response pattern $\mathbf{y}$, $P(\mathbf{Y} = \mathbf{y}, L = c)$, for each of the C latent classes. In general, joint probabilities can be obtained using the equation

$$P(AB) = P(A)P(B|A). \tag{2.5}$$

Thus [recalling that $\gamma_c = P(L = c)$],

$$P(\mathbf{Y} = \mathbf{y}, L = c) = P(L = c)P(\mathbf{Y} = \mathbf{y}|L = c) = \gamma_c \prod_{j=1}^{J} \prod_{r_j=1}^{R_j} \rho_{j,r_j|c}^{I(y_j = r_j)}. \tag{2.6}$$

For instance, in the hypothetical example:

$P(\mathbf{Y} = (\text{Pass, Pass}), L = \text{At Grade Level}) = P(L = \text{At Grade Level}) \times$
$P(\mathbf{Y} = (\text{Pass, Pass})|\text{At Grade Level}) = .7 \times .81 = .57$

and

$P(\mathbf{Y} = (\text{Pass, Pass}), L = \text{Below Grade Level}) = P(L = \text{Below Grade Level}) \times$
$P(\mathbf{Y} = (\text{Pass, Pass})|\text{Below Grade Level}) = .3 \times .04 = .01$.

Once these joint probabilities have been computed, the marginal $P(\mathbf{Y} = \mathbf{y})$ may be found by summing the joint probabilities over the latent classes:

$$P(\mathbf{Y} = \mathbf{y}) = \sum_{c=1}^{C} P(\mathbf{Y} = \mathbf{y}, L = c). \tag{2.7}$$

Substituting Equation 2.6 into Equation 2.7 produces Equation 2.3.

Thus in the hypothetical example, $P(\mathbf{Y} = (\text{Pass, Pass})) = .57 + .01 = .58$. The complete set of response patterns, or cells in the contingency table, appears in Table 2.9. The table also contains the corresponding expected probabilities of each response pattern. (Expected cell counts can be found by multiplying each probability by N.) The reader may wish to compute the remaining three response pattern probabilities as an exercise.

Table 2.9   Response Pattern Probabilities for Hypothetical Example in Table 2.8

Response pattern	Probability
Fail, Fail	.20
Fail, Pass	.11
Pass, Fail	.11
Pass, Pass	.58

2.5.2   The local independence assumption

The latent class models reviewed in this book make few assumptions. There are no assumptions that the indicators of the latent classes or the latent classes themselves are at any level of measurement other than nominal. Because the indicators are categorical, their joint distribution is multinomial. It follows that strict distributional assumptions such as multivariate normality are unnecessary.

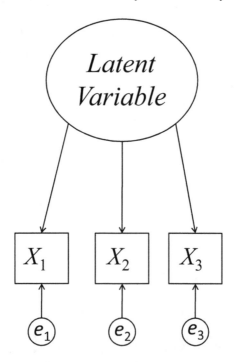

Figure 2.10   Figure 1.1, repeated here for convenience. Latent variable with three observed variables as indicators. This figure illustrates local independence. There are arrows connecting observed variables X_1, X_2, and X_3 to the latent variable but no other arrows connecting any components of the observed variables to each other. This signifies that the three observed variables are related only through the latent variable.

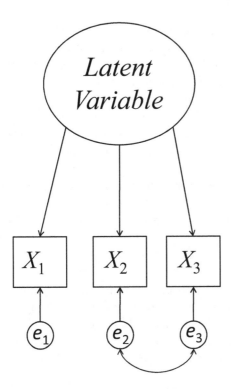

Figure 2.11 Latent variable with three observed variables as indicators. This figure illustrates a violation of local independence. Observed variables X_2 and X_3 are related to each other not only through the latent variable, but also through their error components (e's).

However, there is one fundamental assumption made by latent class models. The assumption of *local independence* specifies that conditional on the latent variable, the observed variables are independent. In other words, if it were possible to create separate contingency tables for the observed variables corresponding to each latent class, the observed variables would be independent within each of these contingency tables (e.g., McCutcheon, 1987). Because the latent variable is not observed directly, such contingency tables can only be approximated.

Although local independence was not mentioned in Chapter 1, it was illustrated in Figure 1.1, repeated here as Figure 2.10. As discussed in Chapter 1, the observed variables are a function of the latent variable and error. In Figure 2.10 the observed variables are connected in only one way: through the latent variable. By contrast, Figure 2.11 illustrates a violation of local independence. In this figure, all of the variables are connected through the latent variable, in the same manner as in Figure 2.10. However, observed variables X_2 and X_3 are also connected through their error components. Thus if the latent variable were conditioned on, X_2 and X_3 would still be related to each other, because e_2 and e_3 are related.

The implications of the local independence assumption can be seen in Equation 2.3. The subscripts on the ρ parameters indicate that the probability of a response to each variable is conditioned on latent class membership only. Then according to the laws of probability, the joint probability of all the elements making up the y vector for latent class c can be found simply by multiplying the individual ρ parameters corresponding to a particular latent class, as shown in Equation 2.3. Without this assumption, Equation 2.3 would have to be much more complicated, because the responses would have to be conditioned not only on latent class membership, but on each other.

The local independence assumption refers only to conditioning on the latent variable. It does *not* imply that in a data set that is to be analyzed, the observed variables are independent. In fact, it is the relations among the observed variables that are explained by the latent classes. An observed data set is a mixture of all the latent classes. Independence is assumed to hold only within each latent class, which is why it is called "local."

It is possible to lift the local independence assumption in latent class models. Examples of work in this area are Hagenaars (1988), Huang and Bandeen-Roche (2004), Reboussin, Ip, and Wolfson (2008), and Uebersax (1999). Such models are outside the scope of this book. All of the latent class models covered in this volume make the local independence assumption.

2.6 SUGGESTED SUPPLEMENTAL READINGS

An overview of latent class models and a history of developments in the field may be found in Goodman (2002). McCutcheon (1987) provides a very readable brief introduction to LCA. Hagenaars (1993) explains latent class models from a log-linear point of view. Lanza, Flaherty, and Collins (2003) offer an introduction to LCA and LTA aimed at those conducting research in psychology and related fields. Lanza, Collins, Lemmon, and Schafer (2007) show details about how to specify a latent class model and fit the model to empirical data. Hagenaars (1988) discusses latent class models that may include local dependencies between the indicators.

Many empirical examples of LCA can be found in the scientific literature. For example, a latent class approach to diagnosis in a medical setting is discussed in Uebersax and Grove (1990) and Rindskopf and Rindskopf (1986). In psychiatry, depression (e.g., Kendler et al., 1996) and substance abuse (e.g., Chung and Martin, 2005) subtypes have been explored. LCA has also been applied to education data (e.g., Aitkin, Anderson, and Hinde, 1981; Dayton, 1991), in child development (Nylund, Bellmore, Nishina, and Graham, 2007), and in marketing research (e.g., Wedel and DeSarbo, 1995).

2.7 POINTS TO REMEMBER

• In the latent class model an individual's observed responses are determined by a combination of the individual's latent class and random error.

• There may be both quantitative and qualitative differences among latent classes.

• LCA can help an investigator to understand a large, complex contingency table.

• The latent class prevalences are the probabilities of membership in each latent class. They are represented by the Greek letter γ (gamma). Because the latent classes are mutually exclusive and exhaustive, they sum to 1.

• The item-response probabilities are the probability of a particular observed response on a particular variable conditional on latent class membership. They are represented by the Greek letter ρ (rho). With each latent class, the item-response probabilities for the set of possible responses for a single item sum to 1.

• The item-response probabilities provide the basis for interpretation and labeling of the latent classes.

• LCA makes the assumption of local independence, which states that conditional on latent class, observed variables are independent. Because the full data set is a mixture of several latent classes, this assumption does not imply that the observed variables are independent in the full sample.

2.8 WHAT'S NEXT

In this chapter we introduced the basics of the latent class model and briefly presented two examples of LCAs conducted on empirical data. This provided a little exposure to interpretation of the item-response probabilities. We demonstrated that the interpretation of LCA rests on these parameters. Because they are so important, in the next chapter we present a detailed exploration of item-response probabilities. In particular, we focus on what the item-response probabilities say about the relations between the latent variable and the observed variables.

CHAPTER 3

THE RELATION BETWEEN
THE LATENT VARIABLE AND
ITS INDICATORS

3.1 OVERVIEW

Every investigator who undertakes LCA has to ponder a set of item-response probabilities in order to interpret them and thereby draw scientific conclusions. Because the item-response probabilities are so critical to the interpretation of LCA, we devote this chapter to how to determine what individual item-response probabilities and the overall pattern of item-response probabilities reveal about the relation between the latent variable and the observed indicator variables.

We begin by discussing LCA as a measurement model and explore the role of latent class membership in determining observed responses. We show how the item-response probabilities associated with an observed variable reveal the strength of its relation with the latent variable. This is followed by an exploration of homogeneity and latent class separation, which are characteristics of the pattern of item-response probabilities taken as a whole. Next we look at the latent class model from a different perspective and consider how to evaluate the precision with which the observed variables measure the latent variable. The chapter ends with a discussion of how investigators may use these ideas in practice.

Most investigators will find that the concepts discussed in this chapter are useful in making decisions about model selection. More formal aspects of model selection, such as hypothesis testing, are discussed in Chapter 4.

3.2   THE LATENT CLASS MEASUREMENT MODEL

3.2.1   Parallels with factor analysis

As mentioned previously, the item-response probabilities associated with an observed variable express its relation with the latent variable. Similarly, in factor analysis an observed variable's factor loading expresses its relation with the factor. A *factor loading* is the coefficient associated with the regression of the variable on the factor. In orthogonal factor analysis, if a variable has a standardized loading near 1 or − 1 on a particular factor, this reflects a strong relation between the variable and the factor. In other words, this means that the observed response is due largely to the latent variable rather than to error.

Because item-response probabilities are conditional probabilities, not regression coefficients, values of item-response probabilities must be interpreted differently from values of factor loadings. In fact, in general the strength of the relation between a single observed variable and the latent variable cannot be determined by examining a single item-response probability. Instead, it is necessary to examine the pattern of item-response probabilities across all response alternatives for the variable and across all of the latent classes.

An advantage of latent variable models, including factor analysis and LCA, is that they not only estimate the amount of measurement error, they also adjust for measurement error. In LCA, this means that the estimates of latent class prevalences are adjusted for measurement error.

3.2.2   Two criteria for evaluating item-response probabilities for a single variable

There are two criteria that define a strong relation between an observed variable j and a latent variable L. The first criterion is a distribution of $\rho_{j,r_j|c}$ for variable j that varies across the latent classes. The second criterion is an array of item-response probabilities corresponding to variable j that are close to 1 and 0. Each of these criteria is necessary for defining a strong relation; neither alone is sufficient.

3.2.2.1   Criterion 1: Distribution of $\rho_{j,r_j|c}$ varies across latent classes

To see why a distribution of $\rho_{j,r_j|c}$ that varies for different latent classes c is a characteristic of a strong relation between an observed variable and a latent variable, let us turn the question around and consider what it means to say that any two variables are unrelated, in other words, are independent. Recall that the item-response probabilities are conditional probabilities, and that according to the laws of probability independence and lack of independence are reflected in conditional probabilities. If two variables A and B are independent, $P(A|B) = P(A)$; if, instead, there is some relation between variables A and B, then $P(A|B) \neq P(A)$.

Suppose that there is a box of dice, half of which are red and half of which are yellow. Color is the only way in which the dice differ, and all of the dice are fair, so that the probability of any number coming up is 1/6. In this example the probability of any number, say two, coming up is independent of the color of the die: $P(\text{two}|\text{red}) = P(\text{two}|\text{yellow}) = 1/6$. Thus knowing the color of the die does not provide any additional information about the probability of observing a two. Moreover, the marginal probability of rolling a two is the same as the conditional probabilities, $P(\text{two}) = 1/6$. In other words, whether or not the probabilities are conditioned on die color, the probability of a two coming up is still the same, because the outcome of the roll of the die is independent of its color, so $P(\text{two}|\text{red}) = P(\text{two}|\text{yellow}) = P(\text{two})$.

Now consider a different scenario, in which the yellow dice are weighted so that a two comes up more often, say 1/3 of the time. Thus the probability of rolling a two is different depending on the die color; $P(\text{two}|\text{red}) = 1/6$, whereas $P(\text{two}|\text{yellow}) = 1/3$. Unlike the example in the preceding paragraph, here the color of the die does contain some additional information about how likely it is that a roll of the die will produce a two. The marginal probability of rolling a two with a randomly selected die is now different from the conditional probabilities: $P(\text{two}) = (1/2)(1/6) + (1/2)(1/3) = 1/4$. Thus, because die color and outcome are not independent, $P(\text{two}|\text{red}) \neq P(\text{two}|\text{yellow}) \neq P(\text{two})$.

Now let us apply these ideas to the item-response probabilities in LCA. When there is no relation between observed variable j and the latent variable, then which response r_j is provided by an individual does not depend on that individual's latent class membership c. Instead, the item-response probabilities *across all the latent classes* are identical and simply equal to the variable's marginal proportions for each response r_j, denoted P_{j,r_j}. Put more formally, when the observed variable and the latent variable are independent, the probability of a response to observed variable j does not depend on the latent variable, and by definition, $\rho_{j,r_j|c} = P_{j,r_j}$ for all $c = 1, ..., C$. When observed variable j and the latent variable L are not independent, then $\rho_{j,r_j|c} \neq \rho_{j,r_j|c'}$ for some c and c'. An example of statistical independence between observed variables and a latent variable is presented in Section 3.2.3.

3.2.2.2   Criterion 2: Item-response probabilities are close to 0 and 1

To understand why item-response probabilities close to 0 and 1 reflect a strong relation between an observed variable and a latent variable, let us consider the meaning of item-response probabilities equal to 1 and 0.

If an item-response probability associated with a variable is 1 for a particular latent class, this means that conditional on membership in that latent class, a particular response can be determined with certainty. If $\rho_{j,r_j|c} = 1$ for response category $r_j = k$, then $\rho_{j,r_j|c} = 0$ for all $r_j \neq k$. For instance, consider the female pubertal development example from Chapter 2. The latent class solution for this data set, presented in Chapter 2, is repeated for convenience in Table 3.1. If $\rho_{4,\text{As curvy/a little more curvy|Delayed Menstrual Onset}} = 1$, then for the girls in the Delayed Menstrual Onset latent class, it is certain that the response to the fourth question will be "As curvy/a little more curvy," and not either of the other two responses. In our example, this estimate was .87, suggesting that girls in the Delayed Menstrual Onset latent class have a high probability of responding "As curvy/a little more curvy" but this response is not certain.

Table 3.1   Item-Response Probabilities from Four-Latent-Class Model of Female Pubertal Development (Add Health Public-Use-Data, Wave I; $N = 469$. From Table 2.5; repeated here for convenience)

	Delayed Menstrual Onset	Biologically Mature	Visibly Mature	Mature
Latent class prevalences	.24	.29	.19	.29
Item-response probabilities				
Ever had menstrual period				
No	**.64***	.16	.41	.09
Yes	.36	**.84**	**.59**	**.91**
How advanced is your physical development compared to that of other girls your age?				
Look younger than most/some	**.73**	.00	.27	.05
Look about average	.24	**.65**	**.52**	.20
Look older than most/some	.02	.35	.21	**.75**
Compared to grade school, my breasts are...				
Same size/a little bigger	**.85**	**.53**	.03	.10
Somewhat bigger	.15	.47	**.80**	.15
A lot/a whole lot bigger	.00	.00	.17	**.75**
Compared to grade school, my body is...				
As curvy/a little more curvy	**.87**	**.63**	.02	.08
Somewhat more curvy	.09	.30	**.86**	.22
A lot/a whole lot more curvy	.04	.07	.12	**.70**

*Item-response probabilities > .5 in bold to facilitate interpretation.

Table 3.2   Hypothetical Item-Response Probabilities Reflecting Independence of Observed Variables and Latent Variable

	Marginal	Latent Class 1	Latent Class 2	Latent Class 3
Variable 1				
Response category 1	.1	.1	.1	.1
Response category 2	.9	.9	.9	.9
Variable 2				
Response category 1	.5	.5	.5	.5
Response category 2	.5	.5	.5	.5
Variable 3				
Response category 1	.8	.8	.8	.8
Response category 2	.2	.2	.2	.2

Unlike a factor loading of 0, which reflects independence between an observed variable and a factor, an item-response probability of 0 also reflects a high degree of certainty, but not necessarily as much certainty as an item-response probability of 1.0. Suppose that $\rho_{j,r_j|c} = 0$ for $r_j = k$. Then conditional on membership in latent class c, it is certain that the response will *not* be k. If there are only two response alternatives, then the item-response probability for the other response must be 1 (by Equation 2.2). However, when there are more than two response categories, a single item-response probability of 0 does not determine what the probability of responses will be for the remaining response categories. For example, Table 3.1 shows that $\rho_{3,\text{A lot/a whole lot bigger|Delayed Menstrual Onset}} = .00$. Suppose this probability were exactly zero (the empirical solution estimated the probability as small enough to round to .00, but it is not exactly 0). This would mean that those in the Delayed Menstrual Onset latent class are certain not to give the response "A lot/a whole lot bigger" to the third question, but it says nothing about the likelihood of the remaining two responses.

3.2.3   Hypothetical and empirical examples of independence and weak relations

Table 3.2 shows item-response probabilities corresponding to three observed variables that are independent of the latent variable. In all three observed variables the distribution of the item-response probabilities is identical across the latent classes, reflecting this independence. However, the values of the item-response probabilities vary considerably. Variable 1 has item-response probabilities that are close to 0 and 1; variable 2's item-response probabilities are far from 0 and 1; and variable 3's item-response probabilities fall between those of variable 1 and variable 2.

Table 3.2 illustrates that no single value or range of values of item-response probabilities can be said always to be indicative of a strong relation between an indicator variable and a latent class. The item-response probabilities in Table 3.2 cover a wide range of possible values, but in every case the indicator variable and the latent variable are independent. Even item-response probabilities very close to 0 and

Table 3.3 Item-Response Probabilities from Five-Latent-Class Model of Health Risk Behaviors (Youth Risk Behavior Survey, 2005; N = 13,840. From Table 2.7; repeated here for convenience)

	Latent Class				
	Low Risk	Early Experi- menters	Binge Drinkers	Sexual Risk- takers	High Risk
Latent class prevalences	.67	.09	.14	.04	.05
Item-response probabilities		Probability of a Yes response			
Smoked first cigarette before age 13	.04	**.76**[*]	.11	.17	**.64**
Smoked daily for 30 days	.02	.31	.27	.12	**.66**
Has driven when drinking	.01	.15	.42	.11	.45
Had first drink before age 13	.14	**.79**	.21	.39	**.68**
≥ 5 drinks in a row in past 30 days	.08	.48	**.74**	.16	**.79**
Tried marijuana before age 13	.01	.46	.03	.22	**.55**
Used cocaine in life	.00	.07	.19	.03	**.88**
Sniffed glue in life	.06	.22	.19	.04	**.58**
Used meth in life	.00	.02	.10	.01	**.73**
Used ecstasy in life	.00	.06	.11	.06	**.64**
Had sex before age 13	.01	.18	.00	**.81**	.30
Had sex with four or more people	.06	.24	.29	**.83**	**.56**

[*]Item-response probabilities > .5 in bold to facilitate interpretation.

Note. The probability of a "No" response can be calculated by subtracting the item-response probabilities shown above from 1.

1 can occur when an observed and latent variable are independent, if the marginal proportions are correspondingly close to 0 and 1. Although for each of the variables in Table 3.2 the indicator variable and the latent variable are independent, the item-response probabilities are different across the variables because they are equal to the marginal proportions, and the three variables have different marginal proportions. (In practice, fit statistics would not point to the nonsensical model in Table 3.2 as an acceptable fit to the data.)

An illustration of an observed and a latent variable that are weakly related, but not independent, can be found in the example from Chapter 2 involving five latent classes of health risk behaviors. The latent class model for this example was shown in Table 2.7 and is repeated in Table 3.3. The third observed variable listed, "Has driven when drinking," appears to have a relatively weak relation with the health risk latent variable. This variable and the latent variable are not independent; the estimates of item-response probabilities corresponding to this variable are different across the latent classes (ranging from .01 to .45), so there is some degree of relation between this observed variable and the latent variable. Moreover, the relation appears to be meaningful; it makes sense that those in the High Risk and Binge Drinkers latent classes would be more likely than members of the other latent classes to report having driven when drinking, because these are latent classes associated with heavier drinking. However, the relation is not as strong as it is for some other variables for which the array of item-response probabilities contains one or two that are above .8 and others that are closer to zero.

3.2.4 Hypothetical and empirical examples of strong relations

Strong relations between indicator variables and the latent variables can be seen in the hypothetical data in Table 3.4, first seen in Chapter 2. The first criterion is met, because there is variability in the item-response probabilities across the three latent classes. The item-response probabilities for Tasks 1 and 2 are the same for Latent Classes 1 and 2 and different for Latent Class 3. Thus these two variables differentiate Latent Class 3 from Latent Classes 1 and 2. Similarly, the item-response probabilities for Task 3 are different for Latent Class 1 than for Latent Classes 2 and 3. Only Task 3 differentiates Latent Classes 1 and 2. The second criterion is met, because the item-response probabilities of .1 and .9 are close to 0 and 1.

As an empirical example of a strong relation between an observed variable and the latent variable, consider the observed variable "Used cocaine in life" in Table 3.3. The probability of a "Yes" response is .88 conditional on the High Risk latent class, and less than .20 conditional on any of the remaining four latent classes. This reflects a stronger relation between this variable and the latent variable as compared to, say, "Has driven when drinking." It is interesting to compare these item-response probabilities to the marginal proportion of individuals responding "Yes" to this item, which can be found in Table 2.6. These marginal proportions indicate that the overall probability of providing a "Yes" response to the cocaine item is .08, whereas for those in the High Risk latent class the probability of a "Yes" response is .88. This difference is striking, and illustrates how important the cocaine variable is in distinguishing the High Risk latent class from the other four.

Table 3.4 Item-Response Probabilities for a Hypothetical Three-Latent-Class Model

Practical Task	Latent Class 1	Latent Class 2	Latent Class 3
Task 1			
Fail	.1	.1	.9
Pass	.9	.9	.1
Task 2			
Fail	.1	.1	.9
Pass	.9	.9	.1
Task 3			
Fail	.1	.9	.9
Pass	.9	.1	.1

3.3   HOMOGENEITY AND LATENT CLASS SEPARATION

In this section we introduce the concepts of homogeneity and latent class separation, two criteria that can be used to evaluate the overall pattern of item-response probabilities. Both criteria have analogs in factor analysis. The term *saturation* is sometimes used to refer to the absolute value of the factor loadings in a factor analysis (e.g., Velicer, Peacock, and Jackson, 1982). The degree to which the loadings are high on a particular factor is sometimes referred to as that factor's saturation. The LCA concept of the *homogeneity* of a latent class is analogous to factor saturation. A pattern of factor loadings that clearly identifies individual factors is called *simple structure* (McDonald, 1985; Thurstone, 1954). The LCA concept of *latent class separation* is analogous to simple structure.

Below we define perfect homogeneity and latent class separation. Like their analogs in factor analysis, the LCA ideas of perfect homogeneity and latent class separation are concepts, not phenomena that an investigator should expect to see exhibited in empirical data. Instead, they should be considered useful benchmarks against which to evaluate a latent class solution. In general, assessment of homogeneity and latent class separation is meant to be a helpful aid to interpretation.

Before discussing homogeneity and latent class separation, let us recall from Chapter 2 that each observed response pattern $\mathbf{y}$ is a vector made up of responses to each of the J observed variables. As before, let r_j represent the response alternatives to observed variable j, so $\mathbf{y} = (r_1, ..., r_J)$. Also recall from Chapter 2 that $P(\mathbf{Y} = \mathbf{y}|L = c)$ can be obtained by using the following expression:

$$P(\mathbf{Y} = \mathbf{y}|L = c) = \prod_{j=1}^{J} \prod_{r_j=1}^{R_j} \rho_{j,r_j|c}^{I(y_j=r_j)}. \tag{3.1}$$

Equation 3.1 will be important in the discussion of homogeneity and latent class separation that follows.

3.3.1   Homogeneity

When latent class c is highly homogeneous, members of c are likely to provide the same observed response pattern, implying that one response pattern is highly characteristic of latent class c. On the other hand, if the homogeneity of latent class c is low, then no single response pattern stands out as by far the most likely for, or clearly characteristic of, latent class c.

The homogeneity of a latent class is at a maximum when all members of the latent class are certain to provide the same observed response pattern. To be more specific, homogeneity of latent class c is perfect when all $\rho_{j,r_j|c} = 0$ or 1 for all variables j and all response categories r_j. In this case, there will be one and only one response pattern $\mathbf{y}'$ for which $P(\mathbf{Y} = \mathbf{y}'|L = c) = 1$; this will be the response pattern for which $\rho_{j,r_j|c} = 1$ for all of those ρ parameters that correspond to the r_j's making up the response pattern $\mathbf{y}'$ (see Equation 3.1). For all of the remaining response patterns there will be at least one ρ parameter that equals 0, so for these response patterns $P(\mathbf{Y} = \mathbf{y}'|L = c) = 0$. In other words, for a latent class with perfect homogeneity, all individuals in that latent class provide exactly the same pattern of responses to all variables.

In empirical data few item-response probabilities are exactly 0 or 1, and therefore for each latent class $c, P(\mathbf{Y} = \mathbf{y}|L = c) > 0$ for many response patterns $\mathbf{y}$. When homogeneity of latent class c is high, many of the item-response probabilities conditioned on c are close to 0 or 1, and as a result, a single $\mathbf{y}$ stands out as having a much larger probability of occurrence conditional on latent class c than any other. Then a single response pattern characterizes latent class c, even though there may be many others with small, but nonzero, probabilities. On the other hand, if the homogeneity of latent class c is low, then there are item-response probabilities conditioned on c that are not close to 0 or 1. In this case out of the array of response patterns there will be more than one, perhaps even many, that emerge as highly likely for individuals in latent class c.

3.3.2 Latent class separation

In factor analysis, a second desirable quality in a factor loading matrix is simple structure, the elegant concept originally developed by Thurstone (1954). In ideal simple structure the overall pattern of factor loadings is such that each variable has a high loading on one and only factor, so that the factor pattern matrix can be arranged in a block format, clearly indicating which variables load on, or characterize, which factor. Simple structure is desirable because it means that the factors are conceptually distinct. Usually when there is simple structure, it is relatively straightforward to interpret the factors based on the factor loadings.

When a set of item-response probabilities is characterized by good latent class separation, the pattern of item-response probabilities across indicator variables clearly differentiates among the latent classes. When there is a high degree of latent class separation, a response pattern that has a large probability of occurrence conditional on one latent class will have much smaller probabilities of occurrence conditional on any of the other latent classes. Put another way, when there is good latent class separation, a response pattern that emerges as characteristic of (i.e., is highly probable conditional on) a particular latent class will be characteristic of that latent class only, and will not be characteristic of any of the other latent classes.

To state this a bit more formally, perfect latent class separation is defined as follows: For each latent class c' there is one response pattern $\mathbf{y}'$ for which $P(\mathbf{Y} = \mathbf{y}'|L = c') = 1$, and $P(\mathbf{Y} = \mathbf{y}'|L = c) = 0$ for all $c \neq c'$. More generally, when there is a high degree of latent class separation, $P(\mathbf{Y} = \mathbf{y}'|L = c') \gg P(\mathbf{Y} = \mathbf{y}'|L = c)$ for all $c \neq c'$.

3.3.3 Hypothetical examples of homogeneity and latent class separation

Homogeneity and latent class separation can be illustrated with some hypothetical examples. Consider a brief questionnaire to be given to college students who smoke at least occasionally. There are three questions: (1) Do you smoke first thing in the morning? (2) Do you purchase packs of cigarettes exclusively for your own use? and (3) Do you go more than three consecutive days without smoking at all? Suppose that there are three response categories for each item: "Always," "Sometimes," and "Never." Let us consider several different hypothetical sets of item-response probabilities corresponding to a three-latent-class model, to illustrate different degrees of homogeneity and latent class separation.

Table 3.5 shows a hypothetical set of item-response probabilities characterized by a high degree of homogeneity and latent class separation. The high level of homogeneity here means that associated with each latent class there is a single, clearly characteristic response pattern that is much more likely to occur than any other. For the first latent class, the characteristic response pattern is (Always, Always, Never); for the second latent class, it is (Sometimes, Always, Sometimes); and for the third latent class, it is (Never, Never, Always). The high level of latent class separation means that no two latent classes have the same characteristic response pattern.

Figure 3.1 displays the item-response probabilities from Table 3.5 in order to illustrate how a high level of homogeneity and latent class separation help make assignment of labels to latent classes straightforward. Individuals in the first latent class, and only individuals in the first latent class, are very likely to respond that they: always smoke first thing in the morning, always purchase a pack of cigarettes exclusively for their own use, and never go more than three days without smoking.

Table 3.5   Item-Response Probabilities for a Hypothetical Three-Latent-Class Model with High Homogeneity and High Latent Class Separation

Question	Regular Smokers	Low-Level Smokers	Light Intermittent Smokers
Smoke first thing			
Always	.90	.05	.05
Sometimes	.05	.90	.05
Never	.05	.05	.90
Purchase pack for own use			
Always	.90	.90	.05
Sometimes	.05	.05	.05
Never	.05	.05	.90
Three days without smoking			
Always	.05	.05	.90
Sometimes	.05	.90	.05
Never	.90	.05	.05

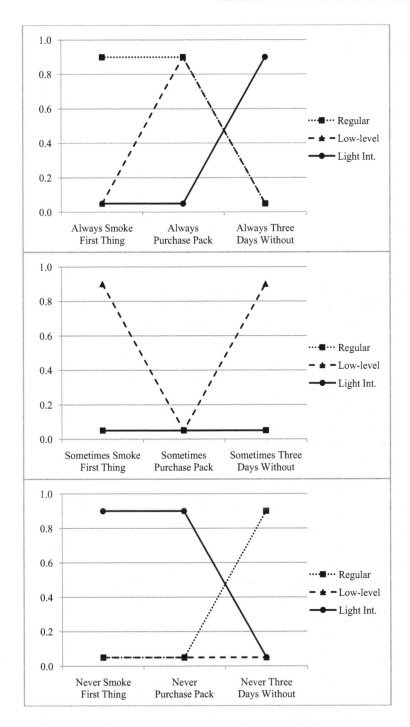

Figure 3.1   Probability of endorsing tobacco use behavior items conditional on latent class membership. Hypothetical data from Table 3.5 exhibit high homogeneity and high latent class separation.

Table 3.6   Item-Response Probabilities for a Hypothetical Three-Latent-Class Model with High Homogeneity and Low Latent Class Separation

Question	Regular Smokers I	Regular Smokers II	Light Intermittent Smokers
Smoke first thing			
Always	.95	.80	.05
Sometimes	.02	.15	.05
Never	.03	.05	.90
Purchase pack for own use			
Always	.90	.80	.05
Sometimes	.05	.10	.05
Never	.05	.10	.90
Three days without smoking			
Always	.03	.05	.90
Sometimes	.02	.15	.05
Never	.95	.80	.05

This suggests that this latent class can be labeled Regular Smokers. Individuals in the second latent class, and only these individuals, are very likely to respond that they: sometimes smoke first thing in the morning, always purchase a pack of cigarettes exclusively for their own use, and sometimes go more than three days without smoking. This latent class can be labeled Low-Level Smokers. Individuals in the third latent class are very likely to respond that they never smoke first thing in the morning, never purchase packs of cigarettes exclusively for their own use, and always go more than three days without smoking. This latent class can be labeled Light Intermittent Smokers.

Table 3.6 shows a different hypothetical set of item-response probabilities characterized by a level of homogeneity comparable to that shown in Table 3.5, but a lesser degree of latent class separation between the first and second latent classes. These item-response probabilities are graphed in Figure 3.2. As Figure 3.2 shows, the first and second latent classes appear to be characterized by the same set of responses: always smoke first thing in the morning, always purchase a pack of cigarettes exclusively for their own use, and never go three days without smoking. The main difference is that the first latent class is somewhat more homogeneous than the second. It would be difficult to come up with labels for these two latent classes that would express a meaningful distinction between them. The same label, Regular Smokers, appears to apply to both latent classes, so all we can do is label them Regular Smokers I and II. (The label Light Intermittent Smokers still applies to the third latent class.)

The hypothetical examples in Tables 3.5 and 3.6 and Figures 3.1 and 3.2 illustrate why it is important to consider homogeneity and latent class separation, which are characteristics of the overall pattern of item-response probabilities, in addition to examining the relation of individual indicator variables and the latent variable. In both Tables 3.5 and 3.6 the individual variables have strong relations with the latent variable. However, the solution in Table 3.5 contains three clearly distinct latent

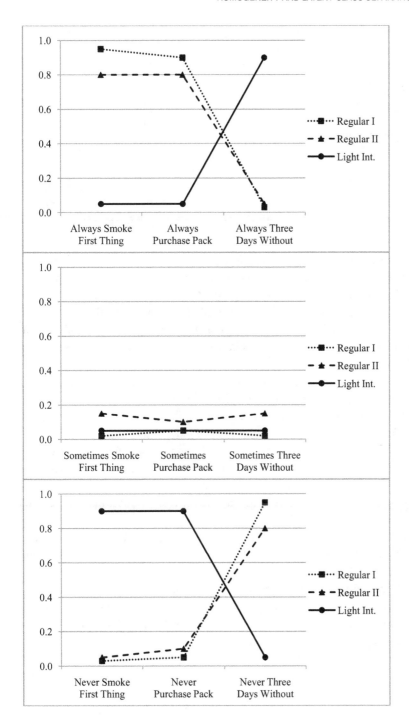

Figure 3.2   Probability of endorsing tobacco use behavior items conditional on latent class membership. Hypothetical data from Table 3.6 exhibit high homogeneity overall and low separation between the Regular I and II latent classes.

Table 3.7   Item-Response Probabilities for a Hypothetical Three-Latent-Class Model with Low Homogeneity and Low Latent Class Separation

Question	Latent Class 1	Latent Class 2	Latent Class 3
Smoke first thing			
Always	.40	.30	.30
Sometimes	.30	.40	.30
Never	.30	.30	.40
Purchase pack for own use			
Always	.40	.40	.30
Sometimes	.30	.30	.30
Never	.30	.30	.40
Three days without smoking			
Always	.30	.30	.40
Sometimes	.30	.40	.30
Never	.40	.30	.30

classes, whereas the solution in Table 3.6 contains two latent classes that are nearly indistinguishable. (In fact, in practice an investigator who obtained the three-latent-class solution in Table 3.6 would probably try a two-latent-class solution, reasoning that the first two latent classes could be combined into one.) This distinction between the solutions in Tables 3.5 and 3.6 would be noted only by examining the overall pattern of item-response probabilities. Graphical aids like Figures 3.1 and 3.2 can help with this process.

Table 3.7 shows a third hypothetical set of item-response probabilities, this time with a low degree of both homogeneity and latent class separation. The overall pattern of larger and smaller item-response probabilities in this table is identical to the one in Table 3.5, but the parameters are closer to the middle of the range ($1/R_j = .33$ for each variable in this example), and therefore the differences between the latent classes are not nearly as pronounced. Low homogeneity means that no clearly characteristic response pattern can be identified for the latent classes. This is evident in Figure 3.3. For example, although an individual in Latent Class 1 is most likely to respond that he or she always smokes first thing in the morning, always purchases a pack of cigarettes for his or her own use, and never goes three days without smoking, there are a number of other response patterns that are nearly as likely to occur. Low latent class separation in this case means that the (Always, Always, Never) response pattern has a large probability in Latent Classes 2 and 3 as well as in Latent Class 1. It is hard to know what unique labels can be applied to each of the latent classes. In general, poor homogeneity and latent class separation can make it very difficult to interpret a latent class solution in a meaningful way.

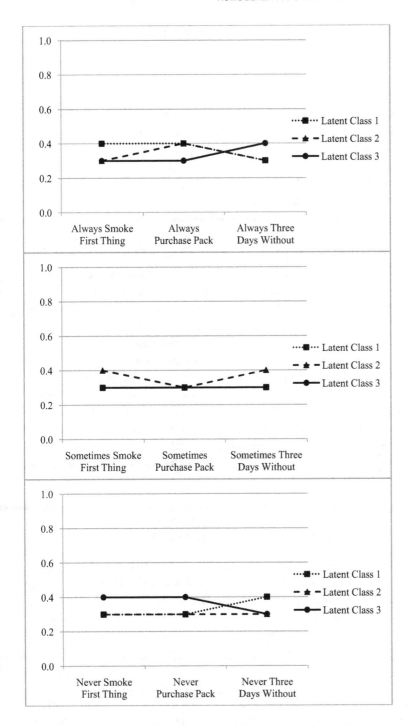

Figure 3.3   Probability of endorsing tobacco use behavior items conditional on latent class membership. Hypothetical data from Table 3.7 exhibit low homogeneity and low latent class separation.

3.3.4 How homogeneity and latent class separation are related

Because homogeneity and latent class separation are based on the same quantities, the item-response probabilities, they are related.

A high degree of latent class separation implies a high degree of homogeneity. This is easiest to see in terms of perfect latent class separation and homogeneity. Let us consider a particular response pattern $\mathbf{y}'$ made up of responses to each of J observed variables. As before we use r_j to represent the response alternatives to observed variable j, so $\mathbf{y}' = (r_1, ..., r_J)$. As explained in Section 3.3.2, in order to have perfect latent class separation $P(\mathbf{Y} = \mathbf{y}'|L = c)$ has to be exactly 1 for one latent class and, by implication, 0 for all the others.

Suppose that for one particular latent class c', $P(\mathbf{Y} = \mathbf{y}'|L = c') = 1$. This implies that $P(\mathbf{Y} = \mathbf{y}'|L = c) = 0$ for all $c \neq c'$. Thus if $P(\mathbf{Y} = \mathbf{y}'|L = c') = 1$, then $\prod_{j=1}^{J} \prod_{r_j=1}^{R_j} \rho_{j,r_j|c'}^{I(y_j'=r_j)} = 1$ (refer to Equation 3.1). It follows that if $P(\mathbf{Y} = \mathbf{y}'|L = c') = 1$, then $\rho_{j,r_j|c'} = 1$ for all of the ρ parameters that are conditional on c' and that correspond to the r_j's making up the response pattern $\mathbf{y}'$. Recall that when $\rho_{j,r_j|c'} = 1$ for all j and all r_j, homogeneity of latent class c' is perfect. This means that perfect latent class separation implies perfect homogeneity.

However, a high degree of homogeneity does not necessarily imply a high degree of latent class separation. Perfect homogeneity for latent class c' requires that $P(\mathbf{Y} = \mathbf{y}'|L = c') = 1$. This does not imply anything about the homogeneity of the remaining latent classes; $P(\mathbf{Y} = \mathbf{y}'|L = c)$ could be anywhere between 0 and 1 for the other latent classes. However, perfect latent class separation requires both that $P(\mathbf{Y} = \mathbf{y}'|L = c') = 1$ for one latent class c' AND that $P(\mathbf{Y} = \mathbf{y}'|L = c) = 0$ for all remaining latent classes $c \neq c'$. Thus it is possible to have high homogeneity but poor latent class separation.

3.3.5 Homogeneity, latent class separation, and the number of response patterns observed

Suppose that a latent class solution were characterized by perfect homogeneity and perfect latent class separation. Then (a) associated with every latent class c' there would be one and only one response pattern $\mathbf{y}'$ for which $P(\mathbf{Y} = \mathbf{y}'|L = c') = 1$, and (b) for this response pattern $\mathbf{y}'$, $P(\mathbf{Y} = \mathbf{y}'|l = c) = 0$ for all $c \neq c'$. In other words, there would be a one-to-one correspondence between latent classes and response patterns, so that there would be exactly as many response patterns observed as latent classes.

For example, if there were perfect homogeneity and latent class separation in Table 3.5, individuals in the Regular Smokers latent class would produce only the (Always, Always, Never) response pattern, individuals in the Low-Level Smokers latent class would produce only the (Sometimes, Always, Sometimes) response pattern, and individuals in the Light Intermittent Smokers latent class would produce only the (Never, Never, Always) response pattern.

Perfect homogeneity and latent class separation can be obtained only when all item-response probabilities are either 0 or 1. However, when all item-response probabilities are either 0 or 1, the model is not really a latent variable model, because the latent classes determine the observed responses with certainty and based on the observed variables, individuals can be assigned to latent classes with certainty. Latent variable models are characterized by some degree of uncertainty in measurement. In the case of latent class models, this uncertainty is reflected in less than perfect homogeneity and latent class separation.

One impact of less than perfect homogeneity and latent class separation is an increase in the number of response patterns that are observed compared to the number that would be observed if there were perfect homogeneity and latent class separation. This increase occurs because now the different individuals within a single latent class may produce different observed response patterns. In Table 3.5, although those in the Regular Smokers latent class are most likely to produce a response pattern of (Always, Always, Never), individuals in this latent class may, with lower probability, respond "Sometimes" or "Never" to the first two items and "Always" or "Sometimes" to the third. Thus an individual who is in the Regular Smokers latent class may produce any of the $3 \times 3 \times 3 = 27$ possible observed response patterns. In fact, the probability of an individual in the Regular Smokers latent class producing the (Always, Always, Never) response pattern is $.90 \times .90 \times .90 = .73$, which means that the probability of the individual *not* producing that response pattern and instead producing one of the other 26 possibilities is $1 - .73 = .27$. Similarly, although the Low-Level Smokers and Light Intermittent Smokers latent classes each is associated with one modal response pattern, members of these latent classes may produce any of the response patterns. In latent class solutions with poorer homogeneity and latent class separation than the example in Table 3.5, the probabilities associated with response patterns that are not characteristic of one of the latent classes tend to be larger.

In short, when there is error, response patterns are observed that would not have been observed if there were no error. The disruptive effect of error in observed responses can result in vastly more expected and observed response patterns than there are latent classes. This is one reason why simply examining response pattern frequencies without modeling measurement error can be an extremely difficult way to gain a sense of the underlying structure of a large contingency table.

3.3.6 Homogeneity and latent class separation in empirical examples

The three hypothetical examples presented above were derived for the purpose of illustrating homogeneity and latent class separation in sharp relief. In empirical data the pattern of homogeneity and latent class separation is rarely either as well established as the one in Table 3.5 or as dismal as the one in Table 3.7. Below we examine the homogeneity and latent class separation exhibited by the two empirical examples first presented in Chapter 2 and discussed earlier in this chapter.

3.3.6.1 The pubertal development example

In the female pubertal development example in Table 3.1, there is fairly good homogeneity. Overall the item-response probabilities tend toward the boundaries of the 0 to 1 range, although homogeneity would be even better if there were more in the 0 to .2 and .8 to 1 range. There are a few areas in the overall pattern in which homogeneity appears a bit weak. For example, consider the Biologically Mature latent class. Although conditional on this latent class the response pattern (2, 2, 1, 1) is most likely, another response pattern, (2, 2, 2, 1), is nearly as likely. This is because two of the response alternatives to the question about breast size have nearly the same probability (.53 and .47).

This empirical example also shows good latent class separation, with the latent classes appearing conceptually distinct overall. For example, only the Delayed Menstrual Onset latent class is characterized by a large probability of response pattern (1, 1, 1, 1). Although the Biologically Mature latent class may be characterized by both the (2, 2, 1, 1) and the (2, 2, 2, 1) response patterns, neither of these patterns has a large probability of occurrence conditional on any other latent class.

3.3.6.2 The health risk behaviors example

Although different from the pubertal development example, the pattern of item-response probabilities associated with the health risk behaviors example shown in Table 3.3 also shows fairly good homogeneity and latent class separation. The Low Risk latent class is particularly homogeneous; a response pattern consisting of "No" responses to all 12 variables clearly characterizes this latent class. Some of the other latent classes are not quite as homogeneous. One example is the Binge Drinkers latent class. In this latent class there is approximately a .42 probability of reporting driving when drinking, which suggests that this latent class is comprised of both individuals who have reported this behavior and individuals who have reported never driving when drinking. For this latent class, the response pattern with the largest conditional probability is one in which there is a "Yes" response to the item "Five or more drinks in a row in past 30 days," and a "No" response to the remaining items. However, there is also a large conditional probability associated with the response pattern which is identical except for a "Yes" response to the question about driving while drinking. The High Risk latent class also shows a relatively low degree of homogeneity; several response patterns have conditional probabilities closer to .5 than to 0 or 1.

Although in this example there is far from perfect homogeneity in several latent classes, latent class separation is good. Some of the latent classes may be characterized by more than one response pattern, but there is no response pattern that appears to be closely associated with more than one latent class. Thus the latent classes are conceptually distinct and can readily be labeled.

3.4 THE PRECISION WITH WHICH THE OBSERVED VARIABLES MEASURE THE LATENT VARIABLE

3.4.1 Why posterior probabilities of latent class membership are of interest

Up to now we have been discussing $P(\mathbf{Y} = \mathbf{y}|L = c)$, the probability of observing response pattern $\mathbf{y}$ conditional on membership in latent class c. In order to discuss the precision with which the set of observed variables measure the latent variable, it is necessary to turn things around and think in terms of the probability of membership in latent class c conditional on response pattern $\mathbf{y}$, $P(L = c|\mathbf{Y} = \mathbf{y})$. This is a very intuitive way of thinking because it reflects the person orientation inherent in LCA. In other words, it enables the investigator to consider each individual's likely latent class membership given his or her observed response pattern. $P(L = c|\mathbf{Y} = \mathbf{y})$ is often called the classification probability or posterior probability. For example, an investigator involved with the latent class analysis of pubertal development shown in Table 3.1 might be interested in each individual's posterior probability of membership in the Mature latent class.

In general, associated with each observed response pattern in LCA there is a nonzero posterior probability of membership in each latent class, except for any latent class c for which $\rho_{j,r_j|c} = 0$ for at least one variable j. When there is a high degree of classification certainty, individuals have a large probability of membership in one and only one latent class and low probabilities in the remaining latent classes. In other words, based on an individual's observed response pattern, it is clear what his or her latent class membership is likely to be. When there is a high degree of classification uncertainty, for many individuals no latent class emerges as by far the most likely given his or her observed response pattern. Some examples of this are reviewed below. We also discuss how the concepts of homogeneity and latent class separation are closely related to classification uncertainty.

Many investigators like the idea of using posterior probabilities to assign individuals to classes, and then to follow this assignment with statistical analyses, such as logistic regression, to predict class membership using new variables. The conceptual appeal of this approach, which is sometimes referred to as classify-analyze, is undeniable. However, the difficulty with such analyses is that they generally do not take into account the uncertainty in classification that is present to some degree in every latent class analysis (e.g., Collins, 2001). This affects any inferences drawn from subsequent analysis. Similar issues exist with respect to estimation of factor scores (e.g., Gorsuch, 1983).

Classification uncertainty is discussed in detail in Sections 3.4.3 and 3.4.4. As will be demonstrated, classification uncertainty varies not only across latent class solutions, but also across individual subjects in a single data set for any particular latent class solution. For this reason we do not recommend using posterior classifications in subsequent analyses, except as a rough exploratory or heuristic device, unless homogeneity and latent class separation are sufficient to assure a high degree of certainty in classification for all individuals, or classification uncertainty is somehow modeled explicitly in the analysis.

In Chapter 6 we discuss a procedure for including covariates in LCA. This procedure does not rely on classification of individuals into latent classes, and does take uncertainty in classification into account in modeling.

3.4.2 Bayes' theorem

Classification probabilities in LCA are obtained using Bayes' theorem. Bayes' theorem is usually stated as

$$P(A|B) = \frac{P(B|A)P(A)}{P(B)}. \tag{3.2}$$

Bayes' theorem is a familiar expression in mathematics, probability, and statistics. In fact, this simple expression is the foundation of Bayesian statistics (e.g., Gelman, Carlin, Stern, and Rubin, 2003).

Let us restate Equation 3.2 in terms that relate more directly to LCA. Substitute the probability of membership in a particular latent class c, $P(L = c)$, for $P(A)$; substitute the probability of a particular response pattern, $P(\mathbf{Y} = \mathbf{y})$, for $P(B)$; and substitute the probability of a particular response pattern conditional on latent class membership, $P(\mathbf{Y} = \mathbf{y}|L = c)$, for $P(B|A)$. Then

$$P(L = c|\mathbf{Y} = \mathbf{y}) = \frac{P(\mathbf{Y} = \mathbf{y}|L = c)P(L = c)}{P(\mathbf{Y} = \mathbf{y})}, \tag{3.3}$$

where $P(L = c|\mathbf{Y} = \mathbf{y})$ represents the posterior probability of membership in Latent Class c conditional on response pattern $\mathbf{y}$.

The estimates of the latent class prevalences and item-response probabilities in a LCA provide the elements needed to obtain the posterior probability of membership in latent class c for individuals who have provided response pattern $\mathbf{y}$. It was shown in Chapter 2 that $P(\mathbf{Y} = \mathbf{y})$ can be obtained as follows:

$$P(\mathbf{Y} = \mathbf{y}) = \sum_{c=1}^{C} \gamma_c \prod_{j=1}^{J} \prod_{r_j=1}^{R_j} \rho_{j,r_j|c}^{I(y_j=r_j)}. \tag{3.4}$$

Then substituting Equations 3.4 and 3.1 into Equation 3.3, and recalling that $\gamma_c = P(L = c)$, produces

$$P(L = c|\mathbf{Y} = \mathbf{y}) = \frac{(\prod_{j=1}^{J} \prod_{r_j=1}^{R_j} \rho_{j,r_j|c}^{I(y_j=r_j)})\gamma_c}{\sum_{c=1}^{C} \gamma_c \prod_{j=1}^{J} \prod_{r_j=1}^{R_j} \rho_{j,r_j|c}^{I(y_j=r_j)}}. \tag{3.5}$$

For example, consider the latent class solution for the female pubertal development data shown in Table 3.1. Suppose that an investigator wishes to obtain the posterior probability that an individual with the response pattern (No, Look about average, Somewhat bigger, Somewhat more curvy) is in the Visibly Mature latent class. This can be obtained as follows:

$P(L = \text{Visibly Mature}|\mathbf{Y} = \text{No, Look about average, Somewhat bigger, Some-}$ what more curvy$) =$

$$\frac{(.19)(.41)(.52)(.80)(.86)}{(.24)(.64)(.24)(.15)(.10) + (.29)(.16)(.65)(.47)(.30) + (.19)(.41)(.52)(.80)(.86) + (.29)(.09)(.20)(.15)(.22)}$$

$= .85$ (within rounding error).

Thus Equation 3.5 can be used to obtain a vector of classification probabilities for each individual based on that individual's observed response pattern and the latent class prevalences and item-response probabilities from any latent class solution fit to the data. This vector will include the probability of membership in each of the C latent classes for that individual.

3.4.3 What homogeneity and latent class separation imply about posterior probabilities and classification uncertainty

Homogeneity and latent class separation are closely related to classification uncertainty. When there is both strong homogeneity and strong latent class separation, the posterior probabilities tend to be large for one single latent class and small for all the others, pointing unambiguously to membership in one latent class for most individuals. In this case there is relatively little classification uncertainty. When either homogeneity or latent class separation is weak, the posterior probabilities tend to be more similar across the latent classes, and there is greater classification uncertainty.

Recall the hypothetical smoking examples used to illustrate high homogeneity and high latent class separation, high homogeneity and low latent class separation, and low homogeneity and low latent class separation in Tables 3.5, 3.6, and 3.7 respectively. By means of Bayes' theorem, Equation 3.2, we can use these examples to demonstrate how homogeneity and latent class separation are closely related to posterior probabilities and classification uncertainty.

Suppose that $\gamma_1 = .2$, $\gamma_2 = .3$, and $\gamma_3 = .5$ in all three examples. Consider the example in Table 3.5, in which homogeneity and latent class separation are both high. Suppose that an individual contributes the (Always, Always, Never) response pattern, which will be abbreviated (A, A, N). Equation 3.5 can be used to compute the posterior probabilities of latent class membership for this individual as follows (all calculations are subject to rounding error):

$$P(L = 1 | \mathbf{Y} = A, A, N)$$

$$= \frac{(.20)(.90)(.90)(.90)}{(.20)(.90)(.90)(.90) + (.30)(.05)(.90)(.05) + (.50)(.05)(.05)(.05)}$$
$$= \quad > .99$$

$$P(L = 2 | \mathbf{Y} = A, A, N)$$

$$= \frac{(.30)(.05)(.90)(.05)}{(.20)(.90)(.90)(.90) + (.30)(.05)(.90)(.05) + (.50)(.05)(.05)(.05)}$$
$$= \quad .01$$

$$P(L = 3 | \mathbf{Y} = A, A, N)$$

$$= \frac{(.50)(.05)(.05)(.05)}{(.20)(.90)(.90)(.90) + (.30)(.05)(.90)(.05) + (.50)(.05)(.05)(.05)}$$
$$\approx \quad .00.$$

Based on the posterior probabilities, an individual who provides this response pattern can be classified into Latent Class 1 with considerable certainty. Now let us compute the posterior probabilities for the same individual but based on the solution in Table 3.6, in which there is still a high degree of homogeneity but less latent class separation (all calculations are subject to rounding error):

$P(L = 1|\mathbf{Y} = \text{A}, \text{A}, \text{N})$

$$= \frac{(.20)(.95)(.90)(.95)}{(.20)(.95)(.90)(.95) + (.30)(.80)(.80)(.80) + (.50)(.05)(.05)(.05)}$$

$$= .51$$

$P(L = 2|\mathbf{Y} = \text{A}, \text{A}, \text{N})$

$$= \frac{(.30)(.80)(.80)(.80)}{(.20)(.95)(.90)(.95) + (.30)(.80)(.80)(.80) + (.50)(.05)(.05)(.05)}$$

$$= .49$$

$P(L = 3|\mathbf{Y} = \text{A}, \text{A}, \text{N})$

$$= \frac{(.50)(.05)(.05)(.05)}{(.20)(.95)(.90)(.95) + (.30)(.80)(.80)(.80) + (.50)(.05)(.05)(.05)}$$

$$\approx .00$$

As might be expected from an examination of Table 3.6, it is quite certain that an individual with this response pattern does not belong in Latent Class 3, but it is much less certain whether the individual should be classified in Latent Class 1 or Latent Class 2. Although Latent Class 1 has the largest probability, there is still a substantial chance that the correct classification would be Latent Class 2.

When there is both low homogeneity and low latent class separation, there is much uncertainty in classification, as can be seen by obtaining the same individual's posterior probabilities based on Table 3.7 (all calculations are subject to rounding error):

$P(L = 1|\mathbf{Y} = \text{A}, \text{A}, \text{N})$

$$= \frac{(.20)(.40)(.40)(.40)}{(.20)(.40)(.40)(.40) + (.30)(.30)(.40)(.30) + (.50)(.30)(.30)(.30)}$$

$$= .35$$

$P(L = 2|\mathbf{Y} = \text{A}, \text{A}, \text{N})$

$$= \frac{(.30)(.30)(.40)(.30)}{(.20)(.40)(.40)(.40) + (.30)(.30)(.40)(.30) + (.50)(.30)(.30)(.30)}$$

$$= .29$$

$P(L = 3 | \mathbf{Y} = A, A, N)$

$$= \frac{(.50)(.30)(.30)(.30)}{(.20)(.40)(.40)(.40) + (.30)(.30)(.40)(.30) + (.50)(.30)(.30)(.30)}$$
$$= \quad .36.$$

It may seem surprising that the largest posterior classification probability for this individual is now associated with Latent Class 3. However, upon reflection this makes sense, because the probabilities obtained through Bayes' theorem are based not only on the item-response probabilities but on the latent class prevalences as well. When there is not much information contained in the item-response probabilities to inform assignment to a latent class, that is, when the observed variables are not closely related to the latent classes, classification relies more upon the latent class prevalences. In fact, if there is no information at all contained in the item-response probabilities, the best option is always to assign the modal latent class. To see why this is so, consider how classification would be done in an imaginary pathological latent class model in which all of the variables are independent of the latent classes. In this case the variables contain no information pertaining to latent class membership. As mentioned above, when there is no relation between the observed and latent variables, $\rho_{j,r_j|c} = P_{j,r_j}$ for all $c = 1, ..., C$. Then

$$P(L = c | \mathbf{Y} = \mathbf{y}) = \frac{\gamma_c \prod_{j=1}^{J} \prod_{r_j=1}^{R_j} P_{j,r_j}^{I(y_j=r_j)}}{\left(\sum_{c=1}^{C} \gamma_c\right) \prod_{j=1}^{J} \prod_{r_j=1}^{R_j} P_{j,r_j}^{I(y_j=r_j)}}$$
$$= \gamma_c$$

(because by Equation 2.1, $\sum_{c=1}^{C} \gamma_c = 1$).

3.4.4 Posterior classification uncertainty even with a high degree of homogeneity and latent class separation

As was demonstrated above, sometimes an individual's vector of posterior latent class membership probabilities points to one latent class with a high degree of certainty, whereas other times there may be a great deal of uncertainty associated with classification. The amount of uncertainty varies across latent class models. Models with high homogeneity and latent class separation are associated with less uncertainty overall in posterior classification than are models with low homogeneity and/or low latent class separation.

However, it is important to realize that uncertainty also varies within latent class models across individuals with different response patterns. For example, the latent class solution depicted in Table 3.5 has extremely high homogeneity and latent class separation, and as was seen above, based on this solution, many individuals can be classified with a high degree of certainty. Yet even in this latent class model there are individuals for whom there is much classification uncertainty. Consider an individual who provides the response pattern (S, S, N). Suppose, as we did above, that $\gamma_1 = .2$, $\gamma_2 = .3$, and $\gamma_3 = .5$. Then the posterior probabilities would be computed as follows (all calculations are subject to rounding error):

$$P(L = 1|\mathbf{Y} = \text{S, S, N})$$

$$= \frac{(.20)(.05)(.05)(.90)}{(.20)(.05)(.05)(.90) + (.30)(.90)(.05)(.05) + (.50)(.05)(.05)(.05)}$$

$$= \quad .38$$

$$P(L = 2|\mathbf{Y} = \text{S, S, N})$$

$$= \frac{(.30)(.90)(.05)(.05)}{(.20)(.05)(.05)(.90) + (.30)(.90)(.05)(.05) + (.50)(.05)(.05)(.05)}$$

$$= \quad .57$$

$$P(L = 3|\mathbf{Y} = \text{S, S, N})$$

$$= \frac{(.50)(.05)(.05)(.05)}{(.20)(.05)(.05)(.90) + (.30)(.90)(.05)(.05) + (.50)(.05)(.05)(.05)}$$

$$= \quad .05.$$

Thus there is considerable uncertainty associated with classifying an individual with response pattern (S, S, N). This is because $P(\mathbf{Y} = \text{S, S, N}|L = c)$ is not large for any of the latent classes. In empirical data, if this solution fit well (model fit is discussed in the next chapter), such a response pattern would be rare—but could occur. In fact, it is nearly inevitable that in any empirical data set at least a few individuals, and usually more than a few, will have response patterns for which posterior classification is very uncertain.

3.5 EXPRESSING THE DEGREE OF UNCERTAINTY: MEAN POSTERIOR PROBABILITIES AND ENTROPY

As described earlier in this chapter, when an individual has a posterior probability close to 1 for a particular latent class (and therefore close to 0 for the other latent classes, as the vector sums to 1 for each individual), there is little error associated with assigning individuals to latent classes based on their maximum posterior probability. As demonstrated above, in many latent class models there will be individuals who can be classified with a high degree of certainty based on their observed response pattern, and also individuals whose observed response pattern does not permit them to be classified with such a high degree of certainty. Because of these individual differences in classification error, it can be informative to examine the vector of posterior probabilities for each individual in a study. This is particularly important if class assignment based on the maximum posterior probability is to be conducted. It can be argued that *when class assignment is to be conducted*, a clear delineation of latent classes, reflected by distinct latent classes and posterior probabilities close to 0 and 1 (i.e., strong homogeneity and latent class separation), is the most important characteristic of a good LCA model.

Even when class assignment is not of interest, however, it can be useful to ex- amine summaries that express overall uncertainty in posterior classification. One straightforward summary is the mean posterior probability of each latent class and the variation about the mean. The mean posterior probability and variation corre- sponding to, for example, Latent Class 1 is computed based on individuals whose maximum posterior probability is for Latent Class 1. This provides a class-specific measure of how well the set of measured items predict membership in each latent class.

Along these same lines, Nagin (2005) proposed a diagnostic called the *odds of correct classification (OCC)* for each latent class, which indicates the improvement over chance in a model's assignment accuracy. An OCC of 1 suggests that group assignment is no better than chance; as the mean posterior probability of membership among individuals assigned to a particular class approaches 1, the OCC increases. Having an OCC of 5 or larger for all latent classes is considered desirable because it indicates a model with good latent class separation (Nagin, 2005, pp. 88–89). One advantage of the OCC diagnostic is that it can be used to identify particular latent classes that are not well measured by the set of observed variables.

Other approaches to summarizing uncertainty in posterior classification have been based on entropy. This approach tends to be coarser because it provides a single- number summary for a particular model. Several entropy-based measures of un- certainty have been proposed. (Although the Merriam-Webster online dictionary defines entropy[2] as "the degree of disorder or uncertainty in a system," measures of entropy tend to be scaled so that larger numbers indicate *less* entropy; in other words, larger values typically indicate less classification error.) One entropy-based measure proposed by Ramaswamy, DeSarbo, Reibstein, and Robinson (1993) is commonly

---

[2]entropy. In *Merriam-Webster Online Dictionary*. Retrieved May 18, 2009, from http://www.merriam-webster.com/dictionary/entropy.

used in LCA. It is designated here as E and is defined as

$$E = 1 - \frac{\sum_{i=1}^{n} \sum_{c=1}^{C} -p_{ic} \log p_{ic}}{n \log C}, \tag{3.6}$$

where p_{ic} is individual i's posterior probability of membership in latent class c, and log refers to the natural log. E ranges between 0 and 1, with larger values indicating better latent class separation (Celeux and Soromenho, 1996). It is useful to remember that E is essentially a weighted average of individuals' posterior probabilities. Even when E is fairly close to 1, it is possible for one or more individuals to have high classification uncertainty.

Latent class assignment error can increase simply as a function of the number of latent classes, so indices like E often decrease as the number of latent classes increases. In other words, class assignment can look better purely by chance in a two-latent-class model than in a comparable model with three or more latent classes. For this reason, entropy-based measures can be a poor tool for model selection. (Tools for model selection are discussed in Chapter 4.)

3.6   POINTS TO REMEMBER

• When an observed variable is independent of the latent variable, the item-response probabilities across all the latent classes are equal to the observed variable's marginal probabilities. When an observed variable is strongly related to the latent variable, the array of item-response probabilities across the latent classes for that variable clearly differentiates the latent classes.

• Homogeneity of a latent class is the extent to which all members of the latent class are likely to provide the same observed responses. Item-response probabilities near 0 and 1 correspond to high homogeneity. Homogeneity is analogous to the factor analysis concept of saturation.

• Latent class separation is the extent to which the overall pattern of item-response probabilities across measured variables clearly differentiates among the latent classes. Latent class separation is analogous to the factor analysis concept of simple structure.

• A high degree of latent class separation implies a high degree of homogeneity. However, a high degree of homogeneity does not imply a high degree of latent class separation.

• Bayes' theorem can be used to obtain posterior classification probabilities for individuals based on their observed response patterns and the parameter estimates from

LCA based on the observed data.

• To the extent that there is a high degree of homogeneity and latent class separation, less uncertainty will be associated with posterior classification. However, the amount of uncertainty varies within a data set for different response patterns, and it is typical for at least a few individuals to provide response patterns that are associated with a large amount of classification uncertainty.

• Mean posterior probabilities, the odds of correct classification, and entropy-based measures are tools that can be used to summarize the degree of classification uncertainty in LCA for a particular data set.

3.7 WHAT'S NEXT

In this chapter and Chapter 2, we emphasized conceptual and statistical groundwork for understanding LCA. In the next chapter we discuss some technical issues related to fitting latent class models, such as estimation, identification, and model selection.

CHAPTER 4

PARAMETER ESTIMATION AND MODEL SELECTION

4.1 OVERVIEW

In this chapter we present some information about how parameters are estimated in LCA, and about how an investigator can evaluate model fit and select among competing models. Although for simplicity we focus on LCA, everything in this chapter applies to LTA as well. Because the technical details have been covered by other authors, we have deliberately kept this chapter as conceptual and nontechnical as possible. Our objective is to provide the reader with practical information that will be useful in analysis of empirical data. Readers who wish more technical information are referred to the literature cited in this chapter and to the list of suggested readings in Section 4.9.

For pedagogical reasons, this chapter covers model fit and model selection before it covers how to identify the ML solution. However, in practice an investigator would usually start by identifying the ML solution for one or more models. This would be followed by evaluating the fit of the models and, if several models are under consideration, selecting the most plausible.

4.2   MAXIMUM LIKELIHOOD ESTIMATION

4.2.1   Estimating model parameters

As we discussed in earlier chapters, there are two types of parameters in the traditional latent class model: the latent class prevalences, represented by γ, and the item-response probabilities, represented by ρ. To fit a latent class model to a data set it is necessary to estimate these parameters based on the data.

Some statistical parameters, such as the familiar mean μ and standard deviation σ, can be estimated simply by solving an equation. This is often called a *closed-form solution*. However, in more complex statistical modeling, such as LCA and LTA, closed-form estimates of model parameters are not available. Instead, an iterative approach to parameter estimation must be taken. In iterative approaches successive sets of parameter estimates are tried using a principled search algorithm. Parameters in LCA are often estimated by means of some version of an iterative procedure called the expectation-maximization (EM) algorithm (Dempster, Laird, and Rubin, 1977), sometimes in combination with another iterative procedure, known as the Newton-Raphson algorithm (e.g., Agresti, 1990) . Both of these procedures search for maximum likelihood (ML) parameter estimates, or, put another way, attempt to maximize the likelihood function. For a given data set and model, the likelihood function expresses the likelihood of the observed empirical data, conditional on the model's parameter estimates (Agresti, 1990). In other words, the ML estimates represent the parameter values for which the data are most likely to be observed. Usually in practice it is more convenient to minimize the log of the likelihood function, ℓ.

Whenever an iterative estimation algorithm is used to obtain ML parameter estimates, it is necessary to establish criteria for stopping the procedure; otherwise, the algorithm could continue indefinitely. Usually, two different criteria are specified. One is the maximum number of iterations the procedure will be allowed to make. The other, more important criterion is a stopping rule that is based on determining when the search is close enough to a set of parameter estimates that maximizes or nearly maximizes the likelihood function.

The stopping rule is usually based on a numerical convergence index and an associated convergence criterion. At some point in the estimation procedure, the iterative estimation algorithm becomes close enough to the theoretical minimum of ℓ that it is appropriate to consider that the minimum has been reached. The convergence index tracks the progress of the iterative algorithm to determine when this has occurred. A common convergence index used in LCA is the maximum absolute deviation (MAD) between the parameter estimates in two successive iterations of the estimation procedure. The MAD associated with a particular iteration is computed by calculating the absolute value of the difference between each parameter estimate at the current iteration and its corresponding estimate at the immediately preceding iteration; the value assigned to MAD for the current iteration is the largest number in this array. The idea behind using MAD as a convergence index is that as the estimation procedure nears the ML values, the change in parameter estimates associated with each new iteration becomes smaller and smaller. (There are some pathological conditions, such as underidentification, discussed in Section 4.4, under which the change in parameter estimates may not always become successively smaller.)

The convergence criterion defines when the change between successive iterations of the estimation procedure has become so small as to be trivial, and therefore the estimation procedure is sufficiently close to the ML solution and can stop iterating. In our work we often use a convergence criterion of MAD $\leq$ .000001. This means that the estimation procedure is stopped right after the iteration in which the maximum difference in any parameter estimate between successive iterations has fallen below .000001.

The estimation procedure iterates until either the convergence criterion or the maximum number of iterations specified has been reached, whichever occurs first. If the convergence criterion is achieved before the maximum number of iterations has been reached, it is said that the estimation procedure has converged. If the maximum number of iterations specified is reached before the convergence criterion is achieved, it is said that the estimation procedure did not converge. Note that whether convergence is achieved is partly determined by the choice of a convergence criterion and the maximum number of iterations. An estimation procedure that does not converge with a particular convergence criterion and maximum number of iterations may converge if either the convergence criterion or maximum number of iterations is increased. However, increasing the convergence criterion essentially establishes a lower threshold to determine that the estimation procedure is close enough to the minimum ℓ, and therefore can result in less precise parameter estimates.

Bayesian estimation provides an alternative to ML estimation of parameters in LCA. Recent work has shown these methods to have excellent performance and flexibility (Chung, Lanza, and Loken, 2008; Garrett and Zeger, 2000; Hoijtink, 1998; Lanza, Collins, Schafer, and Flaherty, 2005), although they can be computationally difficult.

4.2.2 Options for treatment of individual parameters: Parameter restrictions

In previous chapters we discussed unrestricted parameter estimation. A freely estimated parameter in LCA and LTA can take on any value in the range 0 to 1. Most estimation procedures for latent class models give the investigator two options for restricting the estimation of individual parameters. Restricted parameters may be *fixed* or *constrained*. A single model may contain any combination of freely estimated, fixed, and constrained parameters.

A parameter that is fixed to a particular value is not estimated. Before estimation begins, its value must be specified in the range 0 to 1. This fixed value is off-limits to the estimation procedure. When parameters are constrained, they are placed in an equivalence set with other parameters. The estimation of all of the parameters in an equivalence set is constrained to be equal to the same value, which can be any value in the range 0 to 1. There are several reasons, discussed below, why one might wish to restrict the estimation of one or more parameters in a model. The application of parameter restrictions is demonstrated in Section 4.7.

4.2.3 Missing data and estimation

Missing data occur in nearly all empirical data, despite the vigorous efforts of investigators to prevent it. Subjects may overlook some questionnaire items or interview questions, or refuse to respond, or provide unintelligible answers. In a longitudinal study, subjects may be unavailable for one or more occasions of data collections, or they may drop out of the study altogether at some point. Missing data cause two general problems. First, frequently the individuals who provide incomplete data are different from those who provide complete data. For example, respondents who fail to complete a lengthy questionnaire may be poorer readers than those who complete the questionnaire. If adjustments are not made for these differences, results may be biased. Second, if subjects with any missing data on the variables to be included in an analysis are removed from the sample, the sample can become very small. This can lead to estimation problems and poor statistical power.

Three missing-data mechanisms have been identified. These should be considered when deciding how to handle missing data in an analysis (Little and Rubin, 2002; Rubin, 1976; Schafer, 1997; Schafer and Graham, 2002). Data may be missing completely at random (MCAR), missing at random (MAR), or missing not at random (MNAR). When the probability of an individual providing incomplete data does not relate to any variable in the study (either in the analysis or not), the data are MCAR. When this probability depends on variables in the analysis, but not other variables, the data are MAR. Modern missing data procedures exist in most LCA software programs to handle data that are MCAR and MAR (which together are referred to as *ignorable missingness*). For data that are MNAR, missingness is related to some (observed or unobserved) variable that is not included in the analysis. In this case, there exists a systematic reason for providing incomplete responses, but the model cannot adjust parameter estimates for that reason and therefore can produce biased results. In reality, missing data are most likely driven by some combination of all three of these mechanisms. However, the current state of the art of missing data procedures for LCA cannot accommodate MNAR data (which is referred to as *nonignorable missing data*).

When applying statistical models in general, two methods for dealing with missing data in a rigorous way typically are available: full-information maximum likelihood

(FIML) and multiple imputation (MI; Schafer, 1997). FIML is a model-based missing data procedure. In FIML, individuals with complete data and partially complete data are analyzed together, and model estimates are adjusted on the basis of all of the information provided by these individuals. FIML procedures handle data assumed to be ignorable given variables included in the analysis. Most software packages that can fit latent class models employ a FIML approach, and the procedure requires no additional input from the user other than perhaps specifying that missing data occur in the data set and what code is used to denote missing data. However, the current state of the art is that FIML approaches cannot handle missingness when it occurs in grouping variables or covariates in latent class models. (Latent class and latent transition models with grouping variables and covariates are discussed in Chapters 5 through 8.) In short, FIML approaches in LCA handle any ignorable missing data on the indicators of the latent variable, but individuals with missing data on a grouping variable or any covariate in the model are deleted from the analysis. This is analogous to how missing data are handled via FIML in structural equation modeling software.

MI is an extremely general approach to missing data that first became popular in the mid-1990s as user-friendly software options became available. In MI, plausible values are imputed in place of missing values in a data set including individuals with complete and partially complete responses. To account for the uncertainty associated with the imputed values (i.e., because the true values of the data are not known), multiple data sets with randomly different plausible values are created. An important advantage of using MI rather than FIML to handle missing data for latent class modeling is that auxiliary variables (variables that will not be included in the final analysis model but may be related to the probability of missingness) can be included in the imputation model to increase the accuracy of the imputation procedure (e.g., Collins, Schafer, and Kam, 2001). In addition, missing data on covariates can also be handled using this approach. However, the disadvantage is that the latent class model must be fit within each imputed data set, and then results must be combined to obtain final model estimates. Although some aspects of combining results across imputed data sets are straightforward, because of the randomness of imputed values, it is possible that model identification might not be adequate in every data set. Also, if model selection is conducted within each imputed data set, slightly different latent class models might be selected, leaving no logical way to combine results across imputations. A comparison of FIML and MI for handling missing data in statistical models may be found in Collins, Schafer, and Kam (2001).

4.3 MODEL FIT AND MODEL SELECTION

Model selection in LCA can be challenging. The primary decision faced by the investigator is how many latent classes to specify, although other decisions related to parameter restrictions (see Section 4.7) often must be made as well. Fortunately, a variety of tools are available to assist the investigator, including tests of absolute model fit, assessment of relative fit of competing models, and cross-validation. Below we summarize many of the available tools and present examples to demonstrate how we selected the number of latent classes for several examples in this book. In addition to the tools discussed in this section, there are two additional considerations that are critically important when evaluating a model: parsimony and model interpretability.

Parsimony is a philosophical principle stating that all else being equal, simpler models are preferred to more complex models. Here *simpler* means "estimating fewer parameters." According to this principle, statistical models should be no more elaborate, that is, should estimate no more parameters, than is absolutely necessary to represent the data adequately (e.g., Box and Jenkins, 1976, p. 17).

We stated in Chapter 1 that we see statistical models as lenses through which investigators examine their data in order to gain useful insights. It follows that in order to be useful a statistical model must be readily interpretable. Here the investigator's background, experience, and familiarity with prior literature in the area under investigation must be drawn upon to judge whether a particular model is interpretable and provides useful insights that help to move science forward.

Evaluating the fit of a statistical model and choosing from among several competing models is a judgment call. As the reader will see in this chapter and in the remainder of the book, we base our decisions about model evaluation and selection on a combination of statistical criteria, parsimony, and interpretability. As we present empirical examples throughout this book, we have attempted to share our thought processes about model evaluation and selection with the reader. We realize that readers may not always agree with our decisions. Our objective is to provide some exposure to the many considerations involved in selecting a latent class model.

4.3.1 Absolute model fit

Absolute model fit refers to whether a specified latent class model provides an adequate representation of the data in absolute terms, without reference to competing models. Of course, the term *adequate* must be operationalized somehow. In most statistical procedures a test statistic is relied upon to determine whether model fit is adequate.

For example, to test for independence between two categorical variables in a contingency table, a chi-square test statistic can be calculated by comparing the expected cell counts based on the model of independence to the observed cell counts. This test statistic is then compared to the reference chi-square distribution corresponding to the degrees of freedom of the model. By comparing the test statistic to the reference chi-square distribution, it is possible to determine the probability of observing a data set at least as deviant as the observed data from what would be expected under the null hypothesis of independence. If the observed data are very unlikely under the null hypothesis (say, one chance in 100, or $p = .01$), there is strong evidence to reject the null hypothesis of independence. Obtaining a reliable p-value for any test statistic requires knowing the reference distribution of the test statistic under the null hypothesis.

In LCA, as in structural equation modeling, the analyst fits a particular model to the observed data and wishes to test the null hypothesis that this is the population model that produced the observed data. This hypothesis testing is based on the same principles as the test for independence described above, but it has a somewhat different flavor. In the contingency table test for independence, the null hypothesis is a "straw man" that typically does not have any inherent scientific value except as a benchmark and is happily rejected. By contrast, in hypothesis testing in LCA (as, for example, in structural equation modeling) the null hypothesis typically does have some scientific value or interest. Thus here one usually hopes to find a model for which the null hypothesis is *not* rejected.

Because LCA is based on a contingency table, the test statistic used is similar to the one used in standard cross-tabulation tests of independence; the expected cell counts (or expected response pattern proportions) are estimated according to the specified model and estimated parameters, then compared to the observed cell counts (or observed response pattern proportions).

4.3.2 The likelihood-ratio statistic G^2 and its degrees of freedom

The likelihood-ratio statistic G^2 (Agresti, 1990), which is commonly used to assess absolute model fit, that is, to reflect how well a latent class model fits observed data, is calculated as follows. Suppose that there are $j = 1, ..., J$ observed variables, and observed variable j has $r_j = 1, ..., R_j$ response categories. The contingency table formed by cross-tabulating the J variables has $w = 1, ..., W$ cells, where $W = \prod_{j=1}^{J} R_j$. For example, in the pubertal development example reported in Chapter 2, the contingency table contains $W = 2 \times 3 \times 3 \times 3 = 54$ cells. Each of the W cells corresponds to a response pattern, as discussed in Chapters 2 and 3. Let f_w represent the observed frequency of cell w, and let $\hat{f}_w$ represent the expected frequency of cell w according to the model that has been fit. Then

$$G^2 = 2 \sum_{w=1}^{W} f_w \log \left(\frac{f_w}{\hat{f}_w} \right), \qquad (4.1)$$

where log refers to the natural log.

As Agresti (1990, p. 49) stated, "The larger the value of G^2, the more evidence there is against the null hypothesis."

4.3.2.1   Computing the degrees of freedom associated with G^2

The degrees of freedom associated with G^2 are computed as follows. Suppose that P parameters are estimated. In LCA

$$df = W - P - 1. \tag{4.2}$$

The number of estimated parameters P in a standard latent class model is the sum of the number of latent class prevalences estimated and the number of item-response probabilities estimated. (Computation of the number of estimated parameters and degrees of freedom in latent transition models is discussed in Chapter 7.)

Let C represent the number of latent classes in the model. Because $\sum_{c=1}^{C} \gamma_c = 1$, one latent class prevalence can always be obtained by subtraction. Thus when no latent class prevalences are fixed or constrained, the number of latent class prevalences estimated is $C - 1$.

For each latent class, there is an item-response probability corresponding to each response alternative for each variable. Because $\sum_{r_j=1}^{R_j} \rho_{j,r_j|c} = 1$ for a particular variable j and latent class c, one item-response probability per variable for each latent class can always be obtained by subtraction. Thus when no item-response probabilities are fixed or constrained, the number of item-response probabilities estimated is $C \sum_{j=1}^{J} (R_j - 1)$.

Consider the latent class model of female pubertal development reported in Table 2.5. This model has four latent classes, so three latent class prevalences are estimated. There are also $4(1 + 2 + 2 + 2) = 28$ item-response probabilities estimated. Thus $P = 31$ for this model.

When parameter restrictions are imposed in a latent class model, that is, some parameters are fixed or constrained, the number of estimated parameters P is reduced, and the degrees of freedom are increased correspondingly. Fixed parameters are not estimated and therefore do not count toward P. An equivalence set counts as a single estimated parameter, irrespective of the number of parameters that form the set. For example, if 10 item-response probabilities are placed in an equivalence set, the set counts as one estimated parameter.

4.3.2.2   The G^2 and missing data

In LCA, the G^2 statistic must be adjusted when there are missing data. This is because the typical G^2 calculated for incomplete data is comprised of two components: one reflecting lack of model fit and one reflecting the degree to which the data depart from MCAR (against the alternative of MAR; Little and Rubin, 2002). The second component, which can be thought of as a nuisance that inflates the G^2, can be estimated and subtracted from the G^2 to arrive at an adjusted G^2, providing an accurate reflection of model fit when there are missing data (Schafer, 1997, pp. 322–324). Statistical programs for LCA typically perform this adjustment automatically. The adjusted G^2 can then be compared to the reference chi-square distribution corresponding to the degrees of freedom in the model.

4.3.2.3   *Sparseness and assessing model fit*

In theory, a p-value for the likelihood-ratio statistic G^2 can be obtained by comparing the G^2 test statistic to the reference chi-square distribution corresponding to the degrees of freedom in the model. This approach to absolute model fit works well in practice under certain circumstances, but breaks down when the contingency table formed by crossing all indicators of the latent class variable is characterized by sparseness. *Sparseness* refers to the extent to which the average expected cell count is small. Sparseness is a function of the total sample size N and the size of the contingency table W; it is usually expressed as N/W.

Unfortunately, as latent class models become more complex, sparseness quickly becomes an issue. This is particularly true in LTA, where W tends to be large and therefore there can be sparseness even with fairly large sample sizes. Koehler (1986), Koehler and Larntz (1980), and Larntz (1978) showed that when N/W is less than 5, the distribution of the G^2 test statistic is not well approximated by the chi-square, and instead is unknown, making it difficult to test the absolute fit of a model.

Several solutions have been proposed to obtain a reference distribution for the G^2 test statistic. Two different approaches that are conceptually similar are the *parametric bootstrap* (McLachlan and Peel, 2000) and a Bayesian procedure called *posterior predictive checks* (Gelman, Meng, and Stern, 1996; Hoijtink, 1998; Rubin, 1984; Rubin and Stern, 1994). Both approaches are based on generating many random data sets, fitting the model to each random data set and computing the test statistic, and then using the resulting distribution of the test statistic across the random data sets as the reference distribution. The approaches differ primarily in how the random data sets are generated. In the parametric bootstrap, the random data sets are generated based on the parameters estimated from the empirical data. In the posterior predictive checks approach, each random data set is generated based on an independent draw of the parameters from their posterior distributions.

Regardless of which approach is taken to derive the reference distribution, a p-value can be obtained for the likelihood ratio test statistic by determining where the empirically-derived G^2 value falls in that distribution. For example, if the G^2 value falls in the uppermost tail of the reference distribution (e.g., above the 95th percentile, resulting in a p-value of less than .05), it is unlikely that a fit statistic at least as large as the empirically-derived one would be observed under the model corresponding to the null hypothesis. Such a finding would provide evidence to reject the null hypothesis that the estimated model is the population model.

4.3.3 Relative model fit

In general, *relative model fit* refers to deciding which of two or more models represents an optimal balance of fit to a particular data set and parsimony. There are two general approaches to assessing relative model fit. One approach is to conduct a likelihood-ratio difference test. The other is to compare information criteria associated with each model under consideration.

4.3.3.1 The likelihood-ratio difference test

When two (or more) models, say Model A and Model B, are *nested*, the likelihood-ratio difference test can be used to compare the models (with one important exception discussed in Section 4.3.3.2). Model B is nested within Model A if the two models are identical except that the more parsimonious Model B includes some additional parameter restrictions.

The likelihood-ratio difference test provides a formal test of the null hypothesis that the less restrictive model, Model A, and the more restricted model, Model B, fit equally well. As an example, suppose that Model B involves only the addition of the parameter restriction that specifies that the latent class prevalences for Latent Classes 1 and 2 are equal. Then the null hypothesis associated with the likelihood-ratio difference test would be that the model constraining the two latent classes to be equal fits as well as the model that allows them to vary, or, put another way, that the two latent class prevalences are equal. Model B is a simpler, more parsimonious model, so if the hypothesis test is not significant, Model B is preferred. Compared to Model B, Model A always will have a smaller G^2 and fewer df because it estimates more parameters, thereby enabling it to fit the model more closely to the particular data set. The question is whether Model B results in a significant decrement in model fit.

The test statistic is computed as follows:

$$G_\Delta^2 = G_B^2 - G_A^2, \tag{4.3}$$

which is assumed to be distributed as a chi-square with $df = df_B - df_A = P_A - P_B$, where P_A and P_B are the number of parameters estimated in Models A and B, respectively. When G_Δ^2 has relatively few degrees of freedom, its distribution is likely to be approximated reasonably well by the chi-square (e.g., Read and Cressie, 1988). When the df associated with G_Δ^2 is very large, there may be the same distributional problems that arise in connection with the G^2 test statistic for absolute fit.

The difference test is appropriate only when Models A and B are statistically nested, in other words, when one model is a restricted version of the other. If it is not possible to start with Model A and add parameter restrictions to arrive at Model B, the two models are not nested and the difference test is not appropriate. Note that if Models A and B are nested, and Models A and C are nested, this does not imply that Models B and C are nested.

4.3.3.2 An important exception

There is an important exception to the rule that nested models can be compared using the likelihood-ratio difference test. In LCA a fundamental question involves deciding on the appropriate number of latent classes to use to represent a data set. It may seem natural to address this question using a nested model approach, for example, using the likelihood-ratio difference test to compare a model with three latent classes to a model with four latent classes. Unfortunately, it is not valid to conduct a likelihood-ratio difference test between models with different numbers of latent classes, even though such models can be considered nested under some circumstances. The reason for this is that the test would have to be performed by comparing the larger model (with C latent classes) to a model of the same dimensions but where one of the latent class prevalences is fixed to zero. In this case, there is insufficient information to estimate the item-response probabilities corresponding to the empty latent class. This makes the distribution of the test statistic undefined.

However, just as the distribution of an empirically derived G^2 value reflecting overall model fit can be obtained using the parametric bootstrap or the posterior predictive checks distribution, so can the distribution of a likelihood-ratio difference test between two models with different numbers of latent classes. Here, bootstrap samples or samples generated from a draw of parameters from the posterior distribution based on the smaller model (with $C - 1$ latent classes) may be used to estimate the distribution of the difference in the likelihood ratio test statistics for the two competing models. By examining where the empirically based difference falls in this distribution, a p-value can be obtained that tests whether the model with more latent classes provides superior fit.

4.3.3.3 Information criteria

A more common and less computationally intensive approach to relative model fit, and one that is appropriate whether or not models are nested, involves relying on information criteria. Several different information criteria have been proposed as a way to compare the relative balance of model fit and parsimony (i.e., model simplicity) when choosing between competing models.

The Akaike information criterion (AIC; Akaike, 1987) and the Bayesian information criterion (BIC; Schwartz, 1978) are well-known tools for comparing competing models in terms of this balance between fit and parsimony. These criteria are often called penalized fit statistics because they impose a penalty on the G^2. The penalty is a function of the number of parameters estimated in the model, P, and in some cases is also a function of the sample size, N. The equations for each of the information criteria are

$$\text{AIC} = G^2 + 2P \qquad (4.4)$$

$$\text{BIC} = G^2 + [\log(N)]P \qquad (4.5)$$

where log refers to the natural log. The penalty is the rightmost term in each equation. Other information criteria have been proposed as well, including the adjusted BIC (Sclove, 1987) and consistent AIC (CAIC; Bozdogan, 1987). In this book we rely primarily on the AIC and BIC.

For all information criteria, a smaller value represents a more optimal balance of model fit and parsimony; thus, a model with the minimum AIC or BIC might be selected. However, because of the varying penalties associated with each criterion, they often do not identify the same model as optimal. Information criteria are likely to be more useful in ruling out models and narrowing down the set of plausible options than in pointing unambiguously to a single best model.

4.3.4   Cross-validation

Cross-validation is a statistical approach to determining whether an estimated model might be generalizable to other data sets, or whether the parameter estimates are driven to a large extent by particular data collected in a sample. This procedure has been proposed as a model selection tool for structural equation models (Cudeck and Browne, 1983) and for latent class models (Collins, Graham, Long, and Hansen, 1994). In its simplest form, this validation approach involves randomly partitioning the data set into two subsets of data: a training data set and a validation data set. First, the training data set is used to estimate the latent class model. Then the fit of the parameter estimates obtained in that analysis is assessed in the validation data set. If the parameter estimates obtained from the model based on training data provide acceptable model fit when applied to the validation data set, the model validates well, providing support for selection of that particular model.

Double cross-validation, also referred to as twofold cross-validation, is a slightly more rigorous procedure where the data set is partitioned randomly into two halves, data set A and data set B (Collins et al., 1994; Cudeck and Browne, 1983). According to this procedure, the latent class model would be fit to data set A, and then model fit would be assessed by applying those parameter estimates to data set B. Then the same latent class model would be fit to data set B, and model fit assessed by applying those estimates to data set A. In theory, the model that has the best fit for the two validation data sets would be selected, although in practice one model might not emerge as clearly superior.

More generally, k-fold cross-validation involves randomly partitioning the data set into k subsamples. The cross-validation procedure is applied k times, where each time one of the k subsamples is used as the validation data set, with the other $k - 1$ subsamples combined to form the larger training data set. The resulting k model fit indices are then evaluated to find the latent class model that provides the best fit across the different validation data sets.

4.4   FINDING THE ML SOLUTION: MODEL IDENTIFICATION, STARTING VALUES, AND LABEL SWITCHING

4.4.1   Overview of model identification issues

Many estimation algorithms require initial values for parameters in order to "kick off" the estimation. These starting values determine where the algorithm will begin the search for the ML estimates. As explained further below, if the model is well identified, any set of starting values will initiate a search that will culminate in finding the ML estimates. If the model is not well identified, in other words, is underidentified, different sets of starting values may result in different estimates, only one of which will be ML. We discuss how to find ML estimates under these circumstances. When a model is underidentified, a ML solution exists, although it may be difficult to find. In extreme cases, a model may be unidentified, which means that a unique ML solution does not exist. In practice, identification is a continuum along which models may vary widely.

4.4.2   Visualizing identification, underidentification, and unidentification

Consider Figure 4.1, which illustrates a hypothetical likelihood function corresponding to a single generic parameter θ. The ML solution estimate is the value of θ that corresponds to the top of the highest peak in the likelihood function. Figure 4.1 is an example of a likelihood function corresponding to a well-identified model. The function is smooth and unimodal with a clear peak. When the likelihood function resembles the one in Figure 4.1, any starting value will arrive at the ML solution. Some starting values may cause the estimation procedure to take longer than others, because they are farther from the ML solution, but any starting value will initiate a search that will culminate in finding the top of the function's highest peak and the corresponding ML parameter estimates.

The likelihood function in Figure 4.1 can be contrasted with the one in Figure 4.2, which illustrates underidentification. In Figure 4.2 the likelihood function is

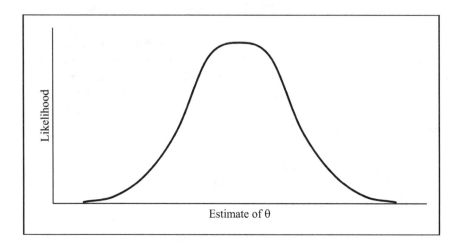

Figure 4.1   Unimodal likelihood function for a single parameter θ, indicative of good identification.

multimodal. There is one global maximum, corresponding to the highest peak, and two local maxima, corresponding to the two lower peaks. A starting value that begins the search near the global maximum is likely to arrive at the ML solution. However, a starting value that happens to begin near one of the local maxima may instead select a solution corresponding to that local maximum. Thus when a model is underidentified, different starting values may arrive at different solutions. However, only one of these is the ML solution. In Section 4.4.4 we discuss a way of finding the ML solution in this situation.

Finally, Figure 4.3 illustrates unidentification. This likelihood function is characterized by a flat region where the value of the likelihood function is maximized across a range of θ's. In this case, there is no unique value of θ that maximizes the likelihood function, and thus a unique ML solution cannot be identified.

In practice, identification is a more complicated issue than it may appear based on Figures 4.1, 4.2, and 4.3. There is a continuum ranging from "identified" to "underidentified" to "unidentified," with infinitely many variations that fall roughly between the examples shown here. Most models involve estimation of many parameters, which means that the likelihood function is multivariate and a starting value is required for each parameter that is estimated. Much of the remainder of Section 4.4 reflects our personal experience with finding the ML solution in empirical data when a latent class model is underidentified, and revising an unidentified model so that a ML solution can be found.

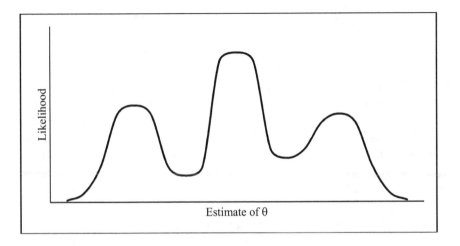

Figure 4.2   Multimodal likelihood function for a single parameter θ, indicative of underidentification.

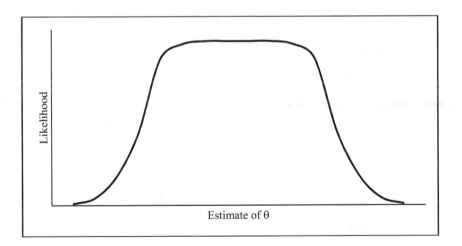

Figure 4.3   Likelihood function for a single parameter θ. This function has a flat region, which suggests that the model being fit is unidentified.

4.4.3 Identification and information

For a latent class model to be identified, the model must have $df \geq 1$. However, this is not sufficient. In fact, it is fairly common for latent class models to be underidentified or even unidentified, even with positive degrees of freedom.

What causes underidentification in LCA when there is at least one degree of freedom? In all statistical models, one necessary condition for identification is that the amount of the "known" information exceeds the amount of "unknown" information. As the ratio of known to unknown information decreases, it can become increasingly difficult to identify the ML solution. In LCA, the unknown information corresponds to what is being estimated, that is, the parameters. Thus, all else being equal, the more parameters are estimated in a model, the greater the chances that there will be identification problems. In particular, as the number of latent classes being fit increases, identification problems tend to increase.

The amount of unknown information can be decreased by imposing parameter restrictions. Fixing a parameter to a pre-specified value instead of estimating it essentially makes that parameter a known. Similarly, unknown information is decreased by placing parameters in an equivalence set. Even though in this case the exact value of the parameters remains unknown, specifying that a set of parameters are all equal to the same value does provide some information. The use of equivalence sets is discussed further in Section 4.7.

The empirical data set to which the model in question is being fit is an important source of information used in parameter estimation. In general, there are three key features of a data set that determine the amount of information it provides. One is the absolute sample size N. All else being equal, a larger N means that there is more information in the data. A second key feature is sparseness, which as mentioned above is expressed in terms of the sample size N in relation to the number of cells W in the contingency table formed by the observed data. All else being equal, a larger N/W means more information. A third key feature is the strength of the relation between the observed variables and the latent variable. All else being equal, if the item-response probabilities in the true underlying model (which is never known) are characterized by a pattern of good homogeneity and latent class separation (see Chapter 3), there is more information in the data. Conversely, there is usually less information when either the sample size or N/W is smaller and when the item-response probabilities of the true underlying model do not show good homogeneity and latent class separation.

4.4.4 How to find the ML solution

When fitting a latent class model investigators should always use multiple sets of starting values to determine whether solutions with different parameter estimates are produced. We recommend beginning by trying a minimum of 10 sets of random starting values and ascertaining whether they consistently converge to the same solution. If they do, it can be concluded that this is the ML solution. If the 10 sets of starting values do not converge to the same solution, we recommend trying many different sets of starting values, say 100 or more, and then examining the distribution of the G^2 or ℓ values (remembering that minimizing the log-likelihood ℓ is equivalent to maximizing the likelihood). Fortunately, in most software packages available for latent class analysis, investigators can readily try different starting values either by using random starting values or providing their own starting values. If the distribution of the G^2 or ℓ values shows that the smallest observed value is also the modal value, this value usually corresponds to the ML solution. This procedure will provide a sense of how confident one can be that the true ML solution has been identified. An empirical example of this approach is presented in Section 4.5.

In our view, an investigator can be most comfortable that the ML solution has been identified if the mode of the distribution of G^2 or ℓ values corresponds to the smallest observed value. If a distribution of G^2 or ℓ is obtained in which there is not a clear mode corresponding to the smallest observed value, sometimes simply increasing the number of sets of starting values used will produce a distribution that points more clearly to the ML solution. However, investigators sometimes find themselves in the frustrating situation of obtaining a distribution of the G^2 or ℓ values across many different sets of starting values that does not point clearly to a ML solution. This suggests that model identification is poor. In this case the only possible avenue is to increase the amount of known information or reduce the amount of unknown information. In theory, known information may be increased by obtaining data on additional subjects, but in practice this is rarely feasible. Therefore, most investigators consider reducing the amount of unknown information by estimating fewer parameters.

Eliminating one or more variables from the contingency table can sometimes help to achieve identification. Reducing the number of variables reduces the number of cells in the contingency table, W, thereby increasing N/W, and also reduces the number of item-response probabilities that must be estimated. However, reducing the number of variables can reduce the amount of information at the same time if the reduction has a negative effect on homogeneity and latent class separation. The strategy also can backfire if conceptually important variables are removed from the analysis. Moreover, a reduction in the size of the contingency table can even result in negative degrees of freedom in some cases.

Another approach is to leave the contingency table unchanged but to change the latent class model being fit so that fewer parameters must be estimated. One way to accomplish this is to apply parameter restrictions. This was discussed briefly above and is discussed further in Section 4.7. Alternatively, a model with fewer latent classes may be fit. This approach can substantially reduce the number of parameters that must be estimated, because it eliminates not only one latent class prevalence but also a corresponding vector of item-response probabilities.

4.4.5   Label switching

When using different sets of starting values and examining the results, it is critical to remember that the ordering of latent classes provided by an estimation algorithm is arbitrary and is determined for a given data set primarily by the starting values. This means, for example, that if a particular set of starting values used in the estimation procedure yields Latent Classes 1, 2, and 3, another set of starting values (applied to the same model and data set) might yield the *identical three latent classes presented in a different order*, so that, for example, what was Latent Class 1 in the first solution would appear as Latent Class 3, Latent Class 2 would appear as Latent Class 1, and Latent Class 3 would appear as Latent Class 2. If the sets of latent classes obtained in the two analyses are identical in terms of their corresponding latent class prevalences and item-response probabilities, the analyses represent the same mathematical solution and will produce the same G^2 and ℓ values. This phenomenon is called *label switching* (Chung, Loken, and Schafer, 2004).

In most latent class analyses, label switching is not intrinsically a problem, but it can be a nuisance. Users who are comparing the results of analyses based on different sets of starting values need to be mindful that label switching can occur, and should look carefully for label switching before concluding that two latent class solutions differ.

4.4.6   User-provided starting values

Some investigators may wish to provide their own starting values in the software they use. When starting values close to the final parameter estimates are provided, convergence can be achieved very quickly. For example, it can be very efficient to retain the parameter estimates from a selected latent class model and use those estimates as starting values when adding covariates to the model (see Chapter 6). Another reason to provide starting values is that in this way the user can have some control over the order in which the latent classes appear on computer output.

Because each parameter in LCA is a probability, starting values must fall between 0 and 1. The boundary values of 0 and 1 should be used only if a parameter estimate is to be fixed to that value, because most estimation procedures will not move away from a starting value of exactly 0 or 1. Importantly, starting values must sum to 1 as appropriate for a particular model. For example, the array of starting values for the latent class prevalences must sum to exactly 1. In addition, for each observed variable the item-response probabilities associated with a particular latent class must sum to exactly 1 across response options.

4.5   EMPIRICAL EXAMPLE OF USING MANY STARTING VALUES: POSITIVE HEALTH BEHAVIORS

Let us consider various health behaviors in a sample of high school seniors in the United States. We are interested in identifying subgroups in the population on the basis of participation in five different positive health behaviors: eating breakfast, eating at least some green vegetables, eating at least some fruit, exercising vigorously, and getting at least seven hours of sleep. The sample consists of $N = 2,065$ high school seniors from the Monitoring the Future 2004 study (Johnston et al., 2005), which provided data on at least one of the five behavior indicators and on both gender and maternal education (the latter two variables are used in analyses reported in Chapters 5 and 6). Each variable was coded 1 for "never/seldom engage in behavior," 2 for "sometimes/most days engage in behavior," and 3 for "nearly every day/every day engages in behavior." Table 4.1 shows the proportion of high school seniors selecting each response category for the five health behaviors. It is interesting to note that the behavior high school seniors are most likely to report never or seldom doing is eating breakfast, and the behavior they are least likely to engage in nearly every day or every day is getting at least seven hours of sleep.

To select the optimal number of latent classes of positive health behaviors, models with one through six latent classes were considered. One hundred different sets of random starting values were used to estimate each of these models. The one-, two-,

Table 4.1   Marginal Response Proportions for Indicators of Positive Health Behavior (Monitoring the Future Data, 2004; $N = 2,065$)

Indicator of Positive Health Behavior	Observed N for Variable	Response Proportion
Eats breakfast	2,058	
Never/seldom		.36
Sometimes/most days		.32
Nearly every day/every day		.32
Eats at least some green vegetables	2,039	
Never/seldom		.18
Sometimes/most days		.49
Nearly every day/every day		.33
Eats at least some fruit	2,037	
Never/seldom		.13
Sometimes/most days		.51
Nearly every day/every day		.36
Exercises vigorously	2,016	
Never/seldom		.25
Sometimes/most days		.42
Nearly every day/every day		.33
Gets at least seven hours sleep	2,042	
Never/seldom		.28
Sometimes/most days		.46
Nearly every day/every day		.26

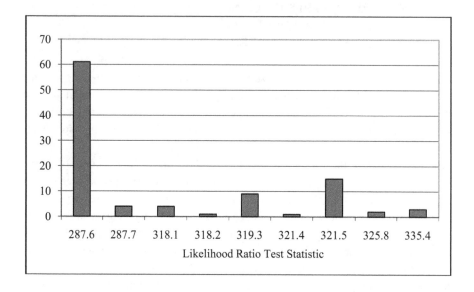

Figure 4.4 Distribution of log-likelihood values for five-latent-class model of positive health behaviors based on 100 random sets of starting values (Monitoring the Future data, 2004; $N = 2,065$).

three-, and four-latent-class models were clearly identified. For the five-latent-class model identification began to break down; that is, the 100 sets of random starting values did not all converge to the same mathematical solution. Identification was much worse for the six-class model. This overall pattern is common. As discussed above, because models with more latent classes usually involve the estimation of more parameters, they tend to have more identification problems.

Figure 4.4 provides considerable insight into identification of the five-latent-class model. This figure shows the frequency distribution of the G^2 likelihood ratio test statistic produced using the 100 sets of starting values. Nine different modes were found, with 15 percent of the sets converging to the solution with $G^2 = 321.5$. Some readers might find the identification of this model to be unacceptable and proceed to reduce the number of "unknowns" in the model by applying parameter restrictions. Others might decide that it would be more appropriate to eliminate the five-latent-class model from consideration. However, given that the estimation procedure converged to the same solution for more than 60 percent of the sets of starting values, and that particular solution also had the lowest G^2 value, we felt confident that this solution represented the ML solution of the five-latent-class model. In other words, we felt that the five-latent-class model was sufficiently well identified and gave this solution consideration when conducting model selection.

4.6 EMPIRICAL EXAMPLES OF SELECTING THE NUMBER OF LATENT CLASSES

Below we present the approach we used to select the number of latent classes used in four of the empirical examples reported in this book.

4.6.1 Positive health behaviors

Our first empirical example of model selection involves choosing the number of latent classes of positive health behaviors using data from the Monitoring the Future 2004 study. In Section 4.5 we showed how we obtained the ML solution for the five-latent-class model using many sets of starting values. Using this same approach we obtained ML solutions for latent class models with a range of numbers of latent classes. Table 4.2 shows the number of parameters estimated, the degrees of freedom, the G^2 statistic, the AIC, the BIC, and the log-likelihood value for these models. Because of the relatively large number of observed variables measuring the latent variable and the number of response categories per variable, the degrees of freedom are fairly large in this example. When the degrees of freedom are large, the reference distribution for the G^2 statistic is not known, so we do not report p-values for tests of model fit. (However, it can be useful to note that the decrease in the G^2 statistic as each latent class is added is fairly substantial even when comparing the four-class model to the five-class model.) Because of the large degrees of freedom, the AIC and BIC were relied on more heavily for model selection.

Figure 4.5 depicts graphically a subset of the information shown in Table 4.2, namely the G^2 statistic, AIC, and BIC for models with one through five latent classes of positive health behaviors. The AIC was lowest for the five-latent-class

Table 4.2 Summary of Information for Selecting Number of Latent Classes of Positive Health Behaviors (Monitoring the Future Data, 2004; $N = 2,065$)

Number of Latent Classes	Number of Parameters Estimated	G^{2*}	df	AIC	BIC	ℓ
1	10	2,532.3	232	2,552.3	2,608.6	−10,655.4
2	21	881.5	221	923.5	1,041.8	−9,830.0
3	32	441.0	210	505.0	685.3	−9,609.8
4	43	356.8	199	442.8	685.0	−9,567.6
5	54	287.6	188	395.6	699.8	−9,533.1
6			Not well identified			

$^*p$-values not reported because the degrees of freedom are too large.

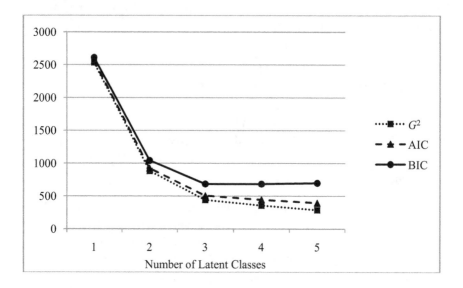

Figure 4.5 G^2, AIC, and BIC for models of positive health behaviors (Monitoring the Future data, 2004; $N = 2,065$).

model, although the BIC pointed to a model with one fewer latent class. Because the information criteria are penalized likelihood-ratio test statistics, often a point of diminishing returns is reached as more and more latent classes (and therefore more parameters) are added. After this point the information criteria may start to increase, even though the G^2 statistic will continue to decrease as the number of latent classes increases. Based primarily on Figure 4.5 and model interpretation, we selected the model with five latent classes. This model is discussed further in Chapter 5.

4.6.2 Past-year delinquency

This empirical example involving data on past-year delinquency was first introduced in Chapter 1. Recall that the subjects are $N = 2,087$ adolescents who participated in the Add Health study (Udry, 2003) and were part of the public-use data set. The adolescents, who were in Grades 10 and 11 (mean age = 16.4) in the 1994–1995 academic year, were asked to respond to questionnaire items asking whether they had engaged in several delinquent behaviors over the previous year.

Models with one through six latent classes were fit to the data. Table 4.3 shows the number of parameters estimated, the G^2 statistic, the degrees of freedom, the p-value associated with the G^2 statistic, the AIC, the BIC, and the log-likelihood value for each

Table 4.3   Summary of Information for Selecting Number of Latent Classes of Past-Year Delinquency (Add Health Public-Use Data, Wave I; $N = 2,087$)

Number of Latent Classes	Number of Parameters Estimated	G^2	df	p-value	AIC	BIC	ℓ
1	6	2,044.0	57	<.0001	2,056.0	2,089.9	−7,021.0
2	13	403.3	50	<.0001	429.3	502.7	−6,200.7
3	20	89.2	43	<.0001	129.2	242.1	−6,043.7
4	27	33.6	36	.58	87.6	240.0	−6,015.8
5				Not well identified			
6				Not well identified			

model. Because the sample size was relatively large ($N = 2,087$) and the degrees of freedom relatively small (less than 60 for all models), we were comfortable comparing the G^2 statistic obtained in each model to a chi-square reference distribution with the corresponding degrees of freedom. For example, the test of the null hypothesis that the model with one latent class fits the data had a p-value < .0001. This provided strong evidence against the null hypothesis, suggesting that a model with more than one latent class was necessary to represent the population. The models with two and three latent classes also had highly significant likelihood-ratio test statistics (p < .0001 for both tests), suggesting that these models did not fit the data well. The four-class model, however, had a G^2 value of 33.6 with 36 degrees of freedom (p = .58), suggesting that the observed data were not unlikely given the specified model. In other words, the four-class model fit the data well. An examination of the AIC and BIC information criteria supported this conclusion as well; because the four-class model had the smallest observed AIC and BIC values, we concluded that this model had the optimal balance of fit and parsimony. The five- and six-latent-class models were not sufficiently identified; thus the AIC and BIC could not be calculated for these models. However, given that the four-class model fit well according to the likelihood-ratio test, there was no need to consider more complex models.

4.6.3   Female pubertal development

This empirical example, which involves selecting a latent class model of female pubertal development, was introduced in Chapter 2. Recall that the subjects are $N = 469$ adolescent girls who participated in Wave I of the Add Health study (Udry, 2003); data were drawn from the public-use data set. The subjects, who were in Grade 7 (mean age = 12.9 years) in the 1994–1995 academic year, were asked to respond to a number of questionnaire items about their physical development.

Models with one through six latent classes were fit to the data. Table 4.4 shows the number of parameters estimated, the G^2 statistic, the degrees of freedom, the p-value, the AIC, the BIC, and the log-likelihood value for each model. Again because

Table 4.4   Summary of Information for Selecting Number of Latent Classes of Female Pubertal Development (Add Health Public-Use Data, Wave I; $N = 469$)

Number of Latent Classes	Number of Parameters Estimated	G^2	df	p-value	AIC	BIC	ℓ
1	7	387.5	46	<.0001	401.5	430.6	−1,777.7
2	15	124.0	38	<.0001	154.0	216.2	−1,645.9
3	23	50.3	30	.01	96.3	191.7	−1,609.0
4	31	25.8	22	.26	87.8	216.5	−1,596.8
5	39	10.2	14	.75	88.2	250.0	−1,589.0
6				Not well identified			

the degrees of freedom were relatively small (less than 50 for all models), issues of sparseness were not likely to affect the distribution of the null model, so we report the p-value for the test of overall fit for each model. The models with one through three latent classes had significant p-values, suggesting that the models did not fit the data well. Both the four- and five-latent-class models, however, provided good fit ($p = .26$ for four latent classes, $p = .75$ for five latent classes), although the six-latent-class model was not well identified. Based on these likelihood-ratio tests, we selected the four-latent-class model.

The G^2 statistic, the AIC and the BIC for each model are plotted in Figure 4.6. In this example the AIC also suggested that the four-latent-class model provided an optimal balance between fit and parsimony, although the BIC suggested that the three-latent-class model was preferable. The reason for the disagreement is that the penalty to the G^2 statistic is simply $2p$ for the AIC (which equals 62 for the four-latent-class model), whereas the penalty for the BIC depends on both the number of parameters and the sample size (in this case the penalty was $[\log(n)]p = 190.7$ for the four-latent-class model). Based on the data presented in Table 4.4 and Figure 4.6, along with an examination of the interpretation of the resultant latent classes, we selected a four-latent-class model of female pubertal development.

4.6.4   Health risk behaviors

The final empirical example involves selecting a model of youth health risk behaviors. This example was introduced in Chapter 2. Recall that the subjects are $N = 13,840$ high school students in the 2005 cohort of the Youth Risk Behavior Survey study (Centers for Disease Control and Prevention, 2004). Subjects responded to 12 questionnaire items about health risk behaviors.

We found the decision about the number of latent classes for this example to be the most challenging of all of those reviewed in this chapter. We fit models with one through seven latent classes to the data. Table 4.5 shows the number of parameters estimated, the G^2 statistic, the degrees of freedom, the AIC, the BIC, and the log-likelihood value for each model. As in Table 4.2, in Table 4.5 we do not report

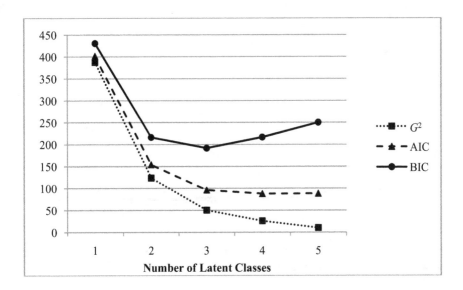

Figure 4.6 G^2, AIC, and BIC for models of female pubertal development (Add Health public-use data, Wave I; $N = 469$).

p-values because there are too many degrees of freedom. Thus we relied primarily on the AIC, the BIC, and model interpretability to decide on the appropriate number of latent classes.

The G^2 statistic, AIC, and BIC for each model are plotted in Figure 4.7. As the figure shows, the AIC and BIC both continued to go down as more latent classes were added. However, there was a leveling off after the five-latent-class solution.

Table 4.5 Summary of Information for Selecting Number of Latent Classes of Health Risk Behaviors (Youth Risk Behavior Survey, 2005; $N = 13,840$)

Number of Latent Classes	Number of Parameters Estimated	G^{2*}	df	AIC	BIC	ℓ
1	12	22,501.7	4,083	22,525.7	22,616.1	−57,634.4
2	25	7,876.9	4,070	7,926.9	8,115.3	−50,322.0
3	38	6,442.2	4,057	6,518.2	6,804.5	−49,604.6
4	51	4,557.3	4,044	4,659.3	5,043.6	−48,662.2
5	64	3,923.4	4,031	4,051.4	4,533.6	−48,345.3
6	77	3,368.1	4,018	3,522.1	4,102.3	−48,067.6
7	90	3,097.8	4,005	3,277.8	3,956.0	−47,932.5

$^*p$-values not reported because the degrees of freedom are too large.

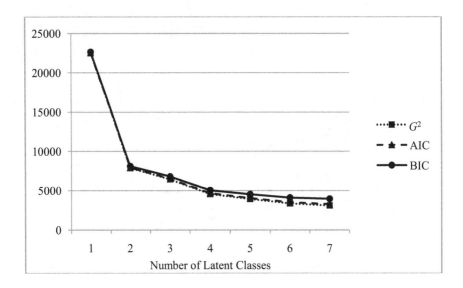

Figure 4.7 G^2, AIC, and BIC for models of health risk behaviors (Youth Risk Behavior Survey, 2005; $N = 13,840$).

Examination of the six- and seven-latent-class solutions showed that the sixth and seventh latent classes were small, and in addition these solutions had poor homogeneity and latent class separation compared to the solution with five latent classes. Based on these considerations we chose the five-latent-class solution. However, this was a judgment call, and it was far from clear-cut. Other investigators might have made a different decision.

4.7   MORE ABOUT PARAMETER RESTRICTIONS

4.7.1   Reasons for using parameter restrictions

There are two primary reasons that an investigator might want to consider the use of parameter restrictions. One reason is to simplify a model being fit, in other words, to reduce the number of parameters being estimated so as to achieve identification. Every parameter restriction imposed on a model simplifies it by reducing the number of parameters that must be estimated, and in turn increasing the degrees of freedom. It is usually possible to achieve model identification by imposing enough parameter restrictions.

The second reason for using parameter restrictions is to express and test hypotheses. An investigator may wish to test an *a priori* hypothesis stemming from theory or prior research suggesting that a particular parameter should equal a specific value. For example, if an investigator is attempting to replicate a previously fit latent class model, it may be interesting to see how well the model and the previously estimated parameters fit in the new data set. This can be done by determining the fit of a model with all of the parameters fixed to the values estimated previously.

More commonly, parameter restrictions are used to set up hypotheses about the equivalence of parameters. In latent class analysis, hypotheses about the equivalence of latent class prevalences may be specified by placing restrictions on the γ parameters, and hypotheses about the equivalence of item-response probabilities may be specified by placing restrictions on the ρ parameters. One important reason for imposing parameter restrictions is to test whether item-response probabilties or latent class prevalences are equivalent across groups. This is discussed further in Chapter 5.

4.7.2   Parameter restrictions and model fit

There is a trade-off between model simplicity and model fit. Every parameter that a model estimates tailors the fit of the model to the observed data. Restricting parameter estimation in a model reduces the overall amount of this tailoring. An equivalence set made up of several parameters is tailored to a particular data set, but because only a single value is estimated, the amount of tailoring is less than would occur if each parameter in the equivalence set were estimated individually. For this reason, imposing parameter restrictions never reduces G^2. The best that can be hoped for is that the parameter restrictions imposed do not increase G^2; in other words, the value to which a parameter is fixed happens to be the parameter's maximum likelihood estimate, and the true values of all the elements of an equivalence set really are identical. If these conditions are met, G^2 will be identical with and without the parameter restrictions. Of course, these conditions are almost never met, so parameter restrictions nearly always result in some increase in G^2. Chapters 5 through 8 contain discussions and demonstrations about testing whether this increase in G^2 is statistically significant.

4.7.3   Using parameter restrictions
to achieve positive degrees of freedom

As noted above, a necessary (but not sufficient) criterion for model identification in LCA is that a model has positive degrees of freedom. Therefore, when a model has degrees of freedom equal to or less than zero, the model must be simplified. One way to simplify the model is to specify fewer latent classes. However, an alternative may be to retain the same number of latent classes and apply one or more parameter restrictions to achieve positive degrees of freedom. Parameter restrictions should be imposed only when the resulting restricted model is plausible, or to set up a hypothesis for falsification. It does not make sense to achieve an identified model at the expense of conceptual meaning. Fortunately, it is frequently possible to apply parameter restrictions in ways that considerably reduce model complexity without sacrificing conceptual appeal. This section provides some hints for accomplishing this.

In general, it is reasonable to apply parameter restrictions in the item-response probabilities when two or more of the probabilities are expected to be nearly equal. For instance, consider the five-latent-class model of health risk behaviors. Table 2.7 shows that for the High Risk latent class, the probabilities of responding "Yes" to both of the questionnaire items about smoking were very similar (.64 and .66). It seems reasonable that constraining these two probabilities to be equal to each other would not substantially alter the interpretation or fit of the five-latent-class model. This type of equality constraint can sometimes make the difference between being able and not being able to fit a conceptually appealing model. It is important to remember that absolute model fit is very likely to worsen when parameter restrictions are imposed. However, the hope is that the cost in model fit, which may not be statistically significant, is worth the improvements in model parsimony achieved by applying the restrictions.

We have found several patterns of restrictions on item-response probabilities to be helpful in practice. To demonstrate this we use the hypothetical model with three binary indicators of performance on tasks and three latent classes presented in Table 3.4 and reproduced here as Table 4.6 for convenience. In this example, the data contain only $2^3 = 8$ possible response patterns, whereas a freely estimated three-latent-class model would involve the estimation of 11 parameters (two latent class prevalences and nine item-response probabilities). Thus, this model would have $8 - 11 - 1 = -4$ degrees of freedom. Parameter restrictions would be necessary to fit a three-latent-class model in this example.

Table 4.6   Item-Response Probabilities for a Hypothetical Three-Latent-Class Model (Table 3.4 repeated for convenience)

Practical Task	Latent Class 1 Masters	Latent Class 2 Intermediate Performers	Latent Class 3 Low Performers
Task 1			
Fail	.1	.1	.9
Pass	.9	.9	.1
Task 2			
Fail	.1	.1	.9
Pass	.9	.9	.1
Task 3			
Fail	.1	.9	.9
Pass	.9	.1	.1

Table 4.7   Type A and Type B Errors for Hypothetical Three-Latent-Class Model in Table 4.6

Practical Task	Latent Class 1 Masters	Latent Class 2 Intermediate Performers	Latent Class 3 Low Performers
Task 1			
Fail	A	A	—
Pass	—	—	B
Task 2			
Fail	A	A	—
Pass	—	—	B
Task 3			
Fail	A	—	—
Pass	—	B	B

Note. A dash (—) corresponds to a response consistent with the label assigned to the latent class.

Based on the item-response probabilities shown in Table 4.6, we might label Latent Class 1 the Masters, Latent Class 2 the Intermediate Performers, and Latent Class 3 the Low Performers. It is possible to consider two types of measurement errors in this example, which we label Type A and Type B. The two types of errors are depicted in Table 4.7. A *Type A error* is the probability of failing a task when the label assigned to the latent class suggests that that group would pass. In the example, a Type A error would occur if an individual who is in the Masters latent class fails Task 1. A *Type B error* is the probability of passing a task when the label assigned to the latent class suggests that that group would fail. In the example, a Type B error would occur if an individual who is in the Low Performers latent class passes Task 1. The item-response probability corresponding to each error response can be considered an estimate of the expected error rate corresponding to that response. For example, Table 4.6 shows that the expected Type A error rate for the Masters latent class, Task 1, is .1, and the expected Type B error rate for the Low Performers latent class, Task 3, is also .1. In Table 4.7 dashes represent the responses that are consistent with the latent class labels. For example, an individual in the Masters latent class would be expected to pass each of the tasks. Remember that when there are two response alternatives, as there are in this hypothetical example, the probability of providing an expected response can be calculated by subtracting the corresponding error rate from 1.

When Type A and Type B errors can be identified, it is often straightforward to arrive at sensible patterns of restrictions for the item-response probabilities. If in this example the item-response probabilities corresponding to Type A error rates are constrained to be equal to each other, and the item-response probabilities corresponding to Type B error rates also are constrained to be equal to each other, a three-latent-class model can be estimated. There would still be two latent class prevalences to estimate, but now there would be only two item-response probabilities to estimate: one for the probability of a Type A error and one for the probability of a Type B error. This would be a total of four parameters estimated, so this model would have $8 - 4 - 1 = 3$ degrees of freedom.

Variants of the pattern of parameter restrictions above might also be worth considering when simplifying a model with negative degrees of freedom. One less restrictive variant might constrain the item-response probabilities corresponding to each type of error to be equal within each variable but allow them to vary across variables. Another variant might constrain the item-response probabilities corresponding to each type of error to be equal within each latent class but allow them to vary across latent classes. A more restrictive variant could involve including all item-response probabilities corresponding to either Type A or Type B errors in a single equivalence set. In this example, such a specification would involve the estimation of only a single item-response probability: the probability of failing when a pass would be expected or passing when a fail would be expected.

This approach to reducing the number of "unknowns" in a latent class model often can allow models to be explored that would not otherwise be estimable. For unconstrained models with negative degrees of freedom, the extent to which such parameter restrictions are plausible cannot be tested empirically because the freely estimated model is not identified. However, in some cases it may be possible to compare the fit of models involving parameter restrictions of varying restrictiveness.

4.8   STANDARD ERRORS

The EM algorithm is the most common approach to estimating parameters in LCA. However, standard errors are not a by-product of this method. Several different approaches for obtaining estimates of standard errors for LCA parameters from the likelihood function have been proposed. All of these likelihood-based methods face the important issue of how to obtain standard errors when one or more parameters are estimated to be on the boundary of the parameter space. Boundary values for all probabilities, including the latent class prevalences and item-response probabilities, are 0 and 1. Unfortunately, the estimation of one or more parameters at a boundary value occurs fairly frequently in LCA. (This problem is even more pervasive in LTA, as transition probabilities also can be estimated at 0 or 1.)

One fairly straightforward approach for estimating standard errors of LCA parameters is to estimate the average cross-product (i.e., the outer product) empirically across individuals in the data set. The cross-product is the first derivative of the likelihood function for each subject. (Technical details may be found in Vermunt and Magidson, 2005, p. 51.) However, this approach yields standard errors of exactly zero for parameter estimates on the boundary, which is not necessarily plausible. It would be unusual to believe that the parameter is on the boundary value with certainty; more likely, there were insufficient data in a particular sample to observe each possible response to the indicators within each latent class.

A better approach, which we will refer to as the *standard* approach, involves taking the inverse of the Hessian (also referred to as the information matrix), which yields the covariance matrix. Standard errors of the LCA parameters are obtained by taking the square root of the values on the diagonal of the covariance matrix. This approach is implemented in numerous software packages and is described in technical detail in Bandeen-Roche, Miglioretti, Zeger, and Rathouz (1997) and Vermunt and Magidson (2005, pp. 50–51). Although this is an appealing method, there are several shortcomings to basing standard errors on the inverse of the Hessian. First, the Hessian cannot be inverted, and therefore estimates of the standard errors cannot be obtained if one or more of the parameters are estimated at a boundary value of the parameter space. In some cases software programs are designed to fix those parameters' standard errors to zero automatically in order to avoid problems with computation. However, as stated above, standard errors of zero are not necessarily plausible. Second, this method is based on the assumption that the underlying model is specified correctly—an assumption that cannot be tested. Assuming that the model is correctly specified, however, and boundary estimates are not an issue, this is an ideal method for obtaining estimates of standard errors for LCA parameters.

Another method, which is often referred to as *robust*, relies on the sandwich estimator (e.g., Huber, 1967; White, 1982). This method relies on both the cross product matrices and the Hessian matrix. This approach is optimal when parts, but not all, of the specified model are correct.

Several Bayesian approaches to estimating standard errors have been applied as well. The two we discuss here provide an attractive alternative to more standard approaches, and resolve the issue of boundary estimates in different ways. One fairly straightforward Bayesian approach is to add a mild smoothing prior (i.e., a Jeffreys prior; Rubin and Schenker, 1987) to each set of item-response probabilities for a particular variable-latent class combination. This technique biases the parameter estimates for the item-response probabilities away from the boundary, just enough to allow standard errors to be estimated. Because parameter estimates on the boundary are most likely caused by sparseness issues, this approach solves the problem that a sample is not large enough to observe, by chance, every possible response to each item within each latent class.

A second Bayesian approach, demonstrated by Lanza, Collins, Schafer, and Flaherty (2005), involves using data augmentation (DA; Tanner and Wong, 1987), a Markov chain Monte Carlo procedure, to obtain standard errors. DA operates essentially like MI for general missing data problems (described in Schafer, 1997), but here the latent variable is treated as a variable with 100 percent missing responses. The latent class variable is imputed multiple times, the parameter estimates and standard errors are calculated within each imputed data set using straightforward formulas, and results are combined across imputations for final parameter estimates and standard errors. Although DA provides a reliable way to obtain standard errors in LCA, the computational burden of the procedure can be heavy.

4.9   SUGGESTED SUPPLEMENTAL READINGS

For more about estimation of latent class model parameters using the EM algorithm the reader is referred to Goodman (1979) and Everitt (1984). More about Bayesian approaches to estimation of latent class model parameters may be found in Chung, Lanza, and Loken (2008) and Lanza, Collins, Schafer, and Flaherty (2005).

Results of simulation studies informing model selection in LCA may be found in Lin and Dayton (1997), Nylund, Asparouhov, and Muthén (2007, 2008), and Rubin and Stern (1994).

Having a sense of how to handle missing data and what the options are can be helpful in conducting LCA and LTA. Graham (2009) provides a readable overview of missing data issues and solutions from a practical perspective. Schafer and Graham (2002) sum up the state of research in missing data. Collins, Schafer, and Kam (2001) compare two general strategies for handling missing data. Schafer (1997) and Little and Rubin (2002) are two standard references in this field.

4.10   POINTS TO REMEMBER

• Estimation of LCA parameters requires initial starting values for parameters. Most software packages for latent class analysis provide an option for random starting values as well as an option for user-provided starting values.

• The ordering of latent classes is arbitrary, and typically determined by the starting values.

• When a particular statistical model and data set provide adequate information to identify the ML estimates of the parameters, the model is considered to be identified. When this is the case, the same solution will be obtained regardless of starting values.

• One necessary but not sufficient) condition that must be met for a latent class model to be identified is that the degrees of freedom must be greater than or equal to 1.

• In general, a latent class model is more likely to be well identified when (1) the sample size N is larger in absolute terms; (2) N/W is larger, where W represents the number of cells in the contingency table formed by the observed data; and (3) there is good homogeneity and latent class separation. A latent class model is less likely to be identified when the sample size or N/W is smaller and when homogeneity or latent class separation are poor.

• It is a good idea to check for underidentification by determining whether different sets of starting values produce different parameter estimates (and therefore different values of ℓ and G^2).

• Absolute model fit refers to whether a specified latent class model provides an adequate representation of the data. One measure of absolute fit is the likelihood-ratio statistic G^2, which reflects the correspondence between observed and expected cell counts in the contingency table made up by crossing all indicators of the latent class variable.

• Relative model fit refers to determining which of several candidate models is optimal. A common approach to relative model fit is to use information criteria, such as the AIC and BIC. These criteria are particularly useful for determining the optimal balance between fit and parsimony when selecting the number of latent classes.

• The parsimony of a latent class model, as well as interpretability of the latent classes, should play a role in model selection.

• Parameter restrictions can involve constraining or fixing parameters. When parameters are constrained, they are placed in an equivalence set with other parameters. The estimation of all of the parameters in an equivalence set is constrained to be equal to the same value. When parameters are fixed, they are not estimated at all. Instead, before estimation begins they are set equal to a user-specified value.

• There are several reasons that an investigator might want to consider the use of parameter restrictions. One reason is to simplify the model being fit, in other words to reduce the number of parameters being estimated, so as to achieve identification. A second reason for using parameter restrictions is to express and test *a priori* hypotheses. When parameter restrictions are imposed, model fit either stays the same or worsens to some degree. The impact of parameter restrictions on model fit can often be assessed using a G^2 difference test.

• When a model has degrees of freedom equal to or less than zero, the model must be simplified. One way to simplify the model is to specify fewer latent classes. However, it may also be plausible to apply one or more parameter restrictions to that model in order to achieve positive degrees of freedom. It often is possible to apply parameter restrictions in ways that considerably reduce model complexity without sacrificing conceptual appeal.

4.11   WHAT'S NEXT

This chapter has provided some technical information that is fundamental to any investigation using LCA and LTA. Some of this material will be revisited in future chapters as needed. We turn next to some advanced topics in LCA: multiple-group LCA (Chapter 5) and then LCA with covariates (Chapter 6).

PART II

ADVANCED
LATENT CLASS ANALYSIS

CHAPTER 5

MULTIPLE-GROUP
LATENT CLASS ANALYSIS

5.1 OVERVIEW

In this chapter we discuss how to approach LCA when there are existing subgroups in the data representing different populations and the investigator wishes to compare certain aspects of the latent class model across these groups (Clogg and Goodman, 1985; McCutcheon, 2002). In the multiple-group models discussed here, the grouping variable is always observed rather than latent. We review how to model and test for group differences in LCA. Group differences may occur in the item-response probabilities and latent class prevalences, or only in the latent class prevalences. We demonstrate how to perform statistical tests for group differences involving comparison of nested models. We also explore some conceptual and interpretational issues.

5.2   INTRODUCTION

Sometimes research questions in LCA involve examining differences between populations, where population membership can be measured directly and the available sample accordingly can be divided into population groups. For example, there may be interest in gender differences, age differences, cohort differences, or differences between experimental conditions.

Group similarities and differences may manifest themselves in a variety of ways. Sometimes the item-response probabilities are identical across groups, which means the latent class prevalences can be compared directly because the interpretation of the latent classes is identical across groups; in other words, any group differences are quantitative, not qualitative. In other situations the item-response probabilities and even the number of latent classes may differ, suggesting that the latent structure itself is different between groups. This reflects group differences that are qualitative at least to some extent. If the latent classes are themselves different between groups, it no longer makes sense simply to compare latent class prevalences directly. Thus whenever there are qualitative group differences, interpretation of the results of LCA becomes more challenging.

The investigator undertaking multiple-group LCA has many decisions to make. Can the same number of latent classes be used to represent each group? Can the item-response probabilities be treated as identical across groups? If not, how different are they? Are the differences relatively minor, so that the interpretation of the latent classes is essentially identical across groups, or do they reflect conceptually important differences in latent structure? Are there significant group differences in the latent class prevalences, and if so, how should they be interpreted? In this chapter we discuss how to approach these questions.

5.3   MULTIPLE-GROUP LCA:
MODEL AND NOTATION

To help introduce multiple-group LCA, let us return to the empirical example of LCA on adolescent delinquency data that was first presented in Table 1.3 and for convenience is repeated here as Table 5.1. In this example there are four latent classes of delinquency: Non-/Mild Delinquents, Verbal Antagonists, Shoplifters, and General Delinquents, measured with six questionnaire items. Recall that the data set involves two cohorts, Grades 10 and 11. In the latent class model shown in Table 5.1, cohort was ignored. Later in this chapter we use multiple-group LCA to examine the following question: To what extent is the latent class model the same or different across the two cohorts? In the present section we keep this example in mind as we discuss the multiple-group LCA model and notation.

Table 5.1   Four-Latent-Class Model of Past-Year Delinquency (Add Health
Public-Use Data, Wave I; $N = 2,087$; Table 1.3 repeated for convenience)

	Latent Class			
Label	Non-/Mild Delinquents	Verbal Antagonists	Shoplifters	General Delinquents
Latent class prevalences	.49	.26	.18	.06
Item-response probabilities corresponding to a Yes* response				
Lied to parents	.33	**.81**†	**.78**	**.89**
Publicly loud/rowdy/unruly	.20	**.82**	**.62**	**1.00**
Damaged property	.01	.25	.25	**.89**
Stolen something from store	.03	.02	**.92**	**.88**
Stolen something worth < $50	.00	.03	**.73**	**.88**
Taken part in group fight	.04	.31	.24	**.64**

*Recoded from original response categories.

†Item-response probabilities > .5 in bold to facilitate interpretation.

As we did in Chapter 2, suppose that there are $j = 1, ..., J$ observed variables measuring the latent classes, and that observed variable j has $r_j = 1, ..., R_j$ response categories. In the delinquency example there are $J = 6$ observed variables, and $R_j = 2$ for each observed variable j. In addition, we introduce a grouping variable V with $q = 1, ..., Q$ groups. In the example V represents cohort, and there are $Q = 2$ groups: the Grade 10 cohort and the Grade 11 cohort.

The contingency table formed by cross-tabulating the J indicator variables and the grouping variable V has $W = Q \prod_{j=1}^{J} R_j$ cells. In the delinquency example, $W = 2 \times 2^6 = 128$. Corresponding to each of the W cells in the contingency table is a complete response pattern, which is a vector of responses to the grouping variable V and the J variables measuring the latent classes, represented by $\mathbf{y} = (q, r_1, ..., r_j, ..., r_J)$. Each response pattern $\mathbf{y}$ in the array corresponding to group q is associated with a probability of occurrence $P(\mathbf{Y} = \mathbf{y}|V = q)$, and within each group q, $\sum P(\mathbf{Y} = \mathbf{y}|V = q) = 1$.

Let L represent the latent variable with $c = 1, ..., C$ latent classes. (To keep the exposition simple, in this section we assume that the number of latent classes is identical across groups.) In our example the latent variable is delinquency, and there are $C = 4$ latent classes. Finally, $I(y_j = r_j)$ is an indicator function that equals 1 when the response to variable $j = r_j$, and equals 0 otherwise. (As mentioned previously, this function is merely a device for picking out the appropriate ρ parameters to multiply together.) Then

$$P(\mathbf{Y} = \mathbf{y}|V = q) = \sum_{c=1}^{C} \gamma_{c|q} \prod_{j=1}^{J} \prod_{r_j=1}^{R_j} \rho_{j,r_j|c,q}^{I(y_j=r_j)}, \qquad (5.1)$$

where $\gamma_{c|q}$ is the probability of membership in latent class c conditional on membership in group q and $\rho_{j,r_j|c,q}$ is the probability of response r_j to observed variable j, conditional on membership in latent class c and group q. In other words, the latent class prevalences and item-response probabilities are conditioned on group membership. In the adolescent delinquency example, this means that a multiple-group LCA can be used to produce estimates of the prevalences of the latent classes and item-response probabilities conditioned on cohort. Separate estimates of latent class prevalences and item-response probabilities can be obtained for Grades 10 and 11, and these estimates can be compared across grades.

5.4 COMPUTING THE NUMBER OF PARAMETERS ESTIMATED

In Chapter 4 we discussed how to compute the number of parameters estimated in standard LCA without a grouping variable. As Equation 5.1 shows, in the multiple-group latent class model the latent class prevalences and item-response probabilities are conditional on the grouping variable. This means that if no parameter restrictions are specified, the number of γ and ρ parameters estimated is multiplied by the number of groups Q. Thus there are $Q(C-1)$ γ's and $QC\sum_{j=1}^{J}(R_j-1)$ ρ's. The number of parameters estimated is reduced if parameter restrictions are specified, as will be seen below.

5.5 EXPRESSING GROUP DIFFERENCES IN THE LCA MODEL

There is maximum similarity across groups when the number of latent classes, the item-response probabilities, and the latent class prevalences are all identical across groups, that is, when the entire latent class model is identical across groups. In this case $\rho_{j,r_j|c,q} = \rho_{j,r_j|c,q'}$ for all items j, response categories r_j, latent classes c, and groups q, q'; $C_q = C_{q'}$ for all groups q, q'; and $\gamma_{c|q} = \gamma_{c|q'}$ for all latent classes c and groups q, q'. Alternatively, different groups may be represented by latent class models with the same number of latent classes and identical item-response probabilities but different latent class prevalences. In this situation $\rho_{j,r_j|c,q} = \rho_{j,r_j|c,q'}$ for all q, q', but $\gamma_{c|q} \neq \gamma_{c|q'}$ for some latent class c and some groups q, q'. In both of the situations above, because the ρ parameters are identical across the groups, the interpretation of the latent classes is identical.

Another possibility is that some or all of the item-response probabilities vary across groups. Here $\rho_{j,r_j|c,q} \neq \rho_{j,r_j|c,q'}$ for at least some variable j, latent class c, and groups q, q'. This means that some groups differ to some extent in the nature of the latent classes. Under these circumstances it would be expected that the latent class prevalences would vary across groups as well, so it would be likely that $\gamma_{c|g} \neq \gamma_{c|q'}$

for at least some latent class c and some groups q, q'. Whether and how the differences in the item-response probabilities affect interpretation of the differences in the latent class prevalences is a matter that requires careful consideration.

Although to keep the exposition simple we did not consider this in Section 5.3, it is possible for the groups to differ in the number of latent classes (i.e., $C_q \neq C_{q'}$). Interestingly, two or more groups may have very similar latent structures even though the number of latent classes is different. For example, suppose that the groups are two cohorts, one younger and one older, and that the younger cohort is represented well by a model with three latent classes, whereas the older one is represented well by a model with four latent classes. It may be that in the older cohort three of the latent classes are identical to those in the younger cohort, and the fourth is an emergent developmentally appropriate latent class. On the other hand, it is also possible for some or all of the latent classes to be very different across groups, whether or not the number of latent classes differs.

5.6 MEASUREMENT INVARIANCE

Whenever two or more groups are to be compared on a latent variable, it is important to establish whether or not the latent variable has the same measurement characteristics in each group. In psychological testing, factor analysis, and structural equation modeling, this concept is known as *measurement invariance*. Millsap and Kwok (2004) define measurement invariance as follows:

> We say that a test fulfills *measurement invariance* across populations when individuals who are identical on the construct being measured, but who are from different populations, have the same probability of achieving any given score on the test. (p. 93)

According to this definition, when there is measurement invariance the distribution of the construct may vary across groups (i.e., populations), but the relation between the construct (i.e., latent variable) and the observed variables is the same. For example, in an achievement testing situation some groups may have higher mean achievement levels and others may have lower mean achievement levels, but a score of, say, 100 has the same meaning, that is, reflects the same degree and kind of achievement, in all groups.

There is a large literature in psychological testing and factor analysis on measurement invariance. Some excellent sources for further reading are Everson, Millsap, and Rodriguez (1991), Horn and McArdle (1992), Meredith (1993), Millsap (1997, 2007), Millsap and Kwok (2004), Vandenberg (2002), Vandenberg and Lance (2000), and Widaman and Reise (1997). Our purpose here is not to review this literature, but to discuss how the concept of measurement invariance can be useful in LCA.

We can begin by extending the definition of measurement invariance provided by Millsap and Kwok (2004) to a latent class framework, as follows: *In LCA, an instrument fulfills measurement invariance across populations when individuals who belong to the same latent class, but who are from different populations, have the same probability of providing any given observed response pattern.*

In factor analysis, measurement invariance is reflected in the factor loadings and intercepts. When there is measurement invariance in the strictest sense, the number of factors and all factor loadings and intercepts are identical across groups. This implies that the interpretation of the factors is identical across groups, and that individuals who have the same value on the latent variable but who are in different groups have identical probabilities of achieving any particular observed score. Similarly, when there is measurement invariance in LCA in the strictest sense, the number of latent classes and all item-response probabilities are identical across groups. This implies that the interpretation of the latent classes is identical across groups, and that individuals in a particular latent class but in different groups have identical probabilities of providing any particular response pattern.

In LCA there are many compelling reasons to hope that measurement invariance across groups can be assumed to hold, or at least to be a reasonable approximation. First, as was mentioned above, when the item-response probabilities are identical across groups, the latent classes have identical interpretations in all groups. Conversely, to the extent that the item-response probabilities differ across groups, then the latent variable itself may be different in the different groups. This makes interpretation of group differences considerably more complicated. Second, a model in which measurement is invariant across groups is simpler, and therefore more parsimonious, than a model in which measurement is allowed to vary across groups (see the discussion of the principle of parsimony in Section 4.3). Third, recall that if there are no parameter restrictions, $QC \sum_{j=1}^{J}(R_j - 1)$ item-response probabilities are estimated. A multiple-group model in which measurement is invariant across groups estimates only $C \sum_{j=1}^{J}(R_j - 1)$ item-response probabilities, or $1/Q$ the number of item-response probabilities as a comparable model with freely estimated item-response probabilities. This reduction in the number of estimated parameters can be large, so when measurement can be considered invariant across groups, often identification problems will be less likely to occur. For these reasons, we recommend constraining item-response probabilities to be equal across groups whenever it is reasonable to do so.

On the other hand, it cannot automatically be assumed that it is appropriate to constrain item-response probabilities to be equal across groups. If the item-response probabilities genuinely differ in important ways across groups, constraining them to be equal will result in a misspecified model, and possibly incorrect scientific conclusions. Of course, the expression "genuinely differ in important ways" does not translate directly into a statistical procedure. Thus an investigator trying to determine whether item-response probabilities can be considered invariant across groups needs to think the problem through carefully. Although statistical procedures are helpful in determining whether measurement invariance holds, ultimately judgment is required on the part of the investigator. In this and many other aspects of empirical applications of latent class analysis, there is no substitute for the investigator's background in the area being studied, knowledge of the data, and thoughtful evaluation of any results.

5.7   ESTABLISHING WHETHER THE NUMBER OF LATENT CLASSES IS IDENTICAL ACROSS GROUPS

In our view, a good place to begin a multiple-group LCA is exploring whether the general latent structure, including the number of latent classes, differs across groups. The most straightforward way to establish whether the number of latent classes is identical across groups is to treat the groups as individual data sets and fit a series of models in each of the groups individually. The decision about whether the number of latent classes is the same across groups must be made carefully and must take interpretability of the solutions into account. Fit indices can be examined in the usual manner, as described in Chapter 4, but there are some additional considerations associated with selecting models when there are multiple groups.

One consideration is the meaning of the principle of parsimony when there are multiple groups. According to the principle of parsimony, when two or more models represent the data approximately equally well, the simplest model should be selected. Ordinarily, this would suggest that all else being equal, the model with the smallest number of latent classes is preferred. However, in multiple-group LCA the principle of parsimony applies not just within the individual groups but to the multiple-group solution as a whole. The principle of parsimony does not necessarily suggest that when individual LCAs are performed in a multiple-group context, model selection should be conducted with a preference for the models with the smallest number of latent classes in each group. The simplest multiple-group model is one in which all groups have the same number of latent classes and measurement invariance holds across all groups for all item-response probabilities.

Another consideration when comparing the relative fit of models with different numbers of latent classes across groups is differences in statistical power. As was discussed in Chapter 4, all else being equal, hypothesis tests are more powerful when N is larger. Thus given the same size discrepancy between the model corresponding to the null hypothesis and the model that generated the data, the p-value associated with the G^2 statistic will be smaller in a group with a larger N as compared to a group with a smaller N.

An additional consideration is that differences in latent class prevalences across groups can lead to group differences in statistical power. If a latent class has a large prevalence in Group A and a small prevalence in Group B, all else being equal there will be less power to detect this latent class in Group B. This may lead to the mistaken conclusion that Group B has fewer latent classes than Group A. This mistaken conclusion may be even more likely if Group B also happens to have a smaller sample size than Group A.

Finally, identification problems may occur if some of the groups have small sample sizes. This may make it difficult or even impossible to fit certain models in each group individually. Often, these identification problems disappear when the groups are combined.

Once the number of latent classes in each group has been established, the groups can be combined in a multiple-group analysis if desired. A multiple-group LCA can be performed if the groups have differing numbers of latent classes, but such an analysis requires particularly careful thought and may present interpretational challenges. In the remainder of this chapter we assume that the number of latent classes is the same in each group.

5.7.1 Empirical example: Adolescent delinquency

We illustrate model selection for multiple-group LCA using the adolescent delinquency example. This example illustrates some of the complexities that can arise in empirical data.

In Chapter 1 it was mentioned that a four-latent-class model had been selected, but the basis for this selection was not described. The first section of Table 5.2 shows goodness-of-fit indices for latent class models with one, two, three, and four latent classes fit to the entire adolescent delinquency sample (note that this information also appeared in Table 4.3). Models with five and six latent classes were attempted, but they were not well identified. Based on the identified models, it appears that the four-latent-class solution represents the data best. Not only is the p-value associated with the G^2 the largest and nonsignificant, the AIC and BIC are the smallest.

Next, a variety of models were fit to each cohort separately to determine whether the two cohorts are best represented by models with the same number of latent classes. The second and third sections of Table 5.2 show the goodness-of-fit indices for Grades 10 and 11, respectively. It appears that the four latent class model fits both the Grades 10 and 11 cohorts better than the one-, two-, or three-latent-class models. Again, the models with five or more latent classes were not well identified.

Therefore, we conclude that it is appropriate to fit both cohorts with latent class models involving four latent classes. Up to this point we have drawn no conclusions about whether any of the item-response probabilities or latent class prevalences are identical across groups. We have simply taken a first step toward establishing the extent of measurement invariance.

Table 5.2   Selecting Number of Latent Classes in Multiple-Group Delinquency
Example (Add Health Public-Use Data, Wave I; $N = 2,087$)

Number of Latent Classes	Number of Parameters Estimated	G^2	df	p-value	AIC	BIC	ℓ
Entire sample							
1	6	2,044.0	57	<.0001	2,056.0	2,089.9	−7,021.0
2	13	403.3	50	<.0001	429.3	502.7	−6,200.7
3	20	89.2	43	.0001	129.2	242.1	−6,043.7
4	27	33.6	36	.58	87.6	240.0	−6,015.8
5				Not well identified			
6				Not well identified			
Grade 10 only ($N = 1,047$)							
1	6	1,082.1	57	<.0001	1,094.1	1,123.9	−3,583.7
2	13	232.7	50	<.0001	258.7	323.1	−3,159.0
3	20	83.3	43	.0002	123.3	222.3	−3,084.3
4	27	41.8	36	.23	95.8	229.6	−3,063.6
5				Not well identified			
6				Not well identified			
Grade 11 only ($N = 1,040$)							
1	6	1,034.0	57	<.0001	1,046.0	1,075.7	−3,433.8
2	13	230.7	50	<.0001	256.7	321.0	−3,032.1
3	20	63.6	43	.02	103.6	202.5	−2,948.6
4	27	38.2	36	.37	92.2	225.7	−2,935.9
5				Not well identified			
6				Not well identified			

5.8 ESTABLISHING INVARIANCE OF ITEM-RESPONSE PROBABILITIES ACROSS GROUPS

Establishing whether the item-response probabilities are invariant across groups requires comparing the fit of two different latent class models. In Model 1, all parameters are allowed to vary across groups. Model 2 is identical to Model 1 except that equivalence sets are defined so that the item-response probabilities are constrained to be equal across groups. If as compared to Model 1, Model 2 provides a significantly poorer fit to the data, the conclusion can be made that restricting the item-response probabilities to be equal across groups is a misspecification, at least for some of the item-response probabilities. On the other hand, if Model 2 fits the data about as well as Model 1, the parameter restrictions in Model 2 are plausible, and it is reasonable to conclude that the item-response probabilities do not differ between the groups. In other words, we can conclude that measurement invariance across groups holds and select the more parsimonious model (Model 2).

This comparison can be made statistically by means of the difference χ^2 test. Let Model 1 be a multiple-group model with C latent classes. In Model 1 all the parameters are free to vary across groups. Let Model 2 be identical to Model 1 except that the item-response probabilities are constrained to be equal across groups; thus,

Models 1 and 2 are nested (see Chapter 4 for a discussion of nested models). If the item-response probabilities are equal across groups, Model 2 will fit the data as well as Model 1. Thus the null hypothesis H_0 can be expressed as follows: "Model 2 fits as well as Model 1" or "Measurement invariance across groups holds." As discussed above, this implies that the parameter restrictions imposed in Model 2 are plausible.

Even when H_0 is true and Models 1 and 2 fit the data equally well, their respective χ^2's are not expected to be equal. This is because compared to Model 1, Model 2 estimates only $1/Q$ as many ρ parameters. Estimation of fewer parameters means more degrees of freedom; in other words, $df_2 > df_1$. The expectation of the χ^2 distribution equals its df, so the expectation of the χ^2 distribution corresponding to Model 2 will always be greater than that corresponding to Model 1. (Another, more intuitive way to look at this is that estimation of fewer parameters means that there are fewer opportunities to fit the solution to the particular data set at hand, so it makes sense that $\chi_2^2 \geq \chi_1^2$.) When H_0 is true, $\chi_2^2 - \chi_1^2$ is itself distributed as a χ^2 with $df = df_2 - df_1$.

Thus a test of H_0 is provided by the likelihood-ratio test $G_\Delta^2 = G_2^2 - G_1^2$, which in theory is distributed as a χ^2 with $df = df_2 - df_1$. A significant hypothesis test, that is, a rejection of H_0, suggests that at least one ρ parameter is different across groups. If H_0 is not rejected, the conclusion is that the ρ parameters can be treated as identical across groups in any subsequent latent class analyses. In addition to the likelihood-ratio statistic G_Δ^2, it is frequently helpful to examine the AIC and BIC in deciding whether or not to conclude that the more restrictive model fits the data sufficiently well. This may be particularly useful when a large number of ρ parameters are involved, as in practice the distribution of the G^2 difference may not be approximated well by the χ^2 when there are many df. As usual, lower AIC and BIC values are associated with the preferred model.

The implication of rejecting H_0 and concluding that Model 1 fits the data better than Model 2 is that one or more of the item-response probabilities, perhaps even all of them, are different across the two groups and need to be estimated separately in any subsequent analyses. Most important, when H_0 is rejected, the investigator must pay careful attention to differences in the interpretation of latent classes in each group. This is discussed further a little later in this chapter. For now, let us examine how to specify the parameter restrictions that are needed to test for measurement invariance.

5.8.1 Specifying parameter restrictions

For the difference χ^2 test to address the question of measurement invariance across groups, Models 1 and 2 must be identical except for whether the item-response probabilities are constrained to be equal across groups. In this section the parameter restriction specifications are illustrated using the adolescent delinquency LCA example. Assuming that the item-response probabilities are all freely estimated in Model

Table 5.3 Parameter Restrictions Specifying Item-Response Probabilities Are Equal Across Grades in Adolescent Delinquency Example

	Latent Class			
	Non-/Mild Delinquents	Verbal Antagonists	Shoplifters	General Delinquents
Grade 10				
Response = Yes*				
Lied to parents	a	b	c	d
Publicly loud/rowdy/unruly	e	f	g	h
Damaged property	i	j	k	l
Stolen something from store	m	n	o	p
Stolen something worth < $50	q	r	s	t
Taken part in group fight	u	v	w	x
Grade 11				
Response = Yes				
Lied to parents	a	b	c	d
Publicly loud/rowdy/unruly	e	f	g	h
Damaged property	i	j	k	l
Stolen something from store	m	n	o	p
Stolen something worth < $50	q	r	s	t
Taken part in group fight	u	v	w	x

*Recoded from original response categories.

Note. Item-response probabilities designated with the same letter form an equivalence set.

1, Table 5.3 illustrates how to specify the equivalence sets in Model 2. The letters are arbitrary, except that parameters designated by the same letter are in the same equivalence set. In Model 1, $QC \sum_{j=1}^{J} (R_j - 1) = 48$ item-response probabilities are estimated. In Model 2, the number of estimated item-response probabilities is reduced to $C \sum_{j=1}^{J} (R_j - 1) = 24$.

In Table 5.3 the only restrictions placed on the item-response probabilities is that they be equal across groups. It is possible to test for measurement invariance across groups when there are additional restrictions placed on the item-response probabilities, as long as these additional restrictions are identical in Model 1 and Model 2. For example, it is not uncommon for Model 1 to be unidentified if all item-response probabilities are freely estimated, particularly when there are many groups and/or many observed variables. Therefore, it may be desirable to add parameter restrictions within each group to achieve identification.

Suppose that in the adolescent delinquency model the item-response probabilities corresponding to the two questions about stealing are to be constrained to be equal to each other, and all other parameters are to be freely estimated. Table 5.4 provides an illustration of how to set up parameter restrictions to test for measurement invariance. In Model 1, all parameters are freely estimated except those corresponding to the two questions about stealing. Equivalence sets are specified so that the two stealing questions are constrained to be equal within Grade 10 and equal within Grade 11. However, these item-response probabilities are not constrained to be equal across

Table 5.4   Parameter Restrictions for Testing Measurement Invariance When Other
Parameter Restrictions Are Present

	Latent Class			
	Non-/Mild Delinquents	Verbal Antagonists	Shoplifters	General Delinquents
Model 1				
Item-response probabilities free to vary across groups				
Grade 10				
Response = Yes				
Lied to parents	*	*	*	*
Publicly loud/rowdy/unruly	*	*	*	*
Damaged property	*	*	*	*
Stolen something from store	m	n	o	p
Stolen something worth < $50	m	n	o	p
Taken part in group fight	*	*	*	p *
Grade 11				
Response = Yes				
Lied to parents	*	*	*	*
Publicly loud/rowdy/unruly	*	*	*	*
Damaged property	*	*	*	*
Stolen something from store	mx	nx	ox	px
Stolen something worth < $50	mx	nx	ox	px
Taken part in group fight	*	*	*	px *
Model 2				
Parameter restrictions consistent with Model 1, plus item-response probabilities constrained equal across groups				
Grade 10				
Response = Yes				
Lied to parents	a	b	c	d
Publicly loud/rowdy/unruly	e	f	g	h
Damaged property	i	j	k	l
Stolen something from store	m	n	o	p
Stolen something worth < $50	m	n	o	p
Taken part in group fight	u	v	w	x
Grade 11				
Response = Yes				
Lied to parents	a	b	c	d
Publicly loud/rowdy/unruly	e	f	g	h
Damaged property	i	j	k	l
Stolen something from store	m	n	o	p
Stolen something worth < $50	m	n	o	p
Taken part in group fight	u	v	w	x

∗ denotes a free parameter.

Note. Item-response probabilities designated with the same letter or pair of letters form an equivalence
set.

cohorts. For example, the item-response probabilities for the two stealing questions
for the Non-/Mild Delinquents latent class are designated with the letters "m" in
Grade 10 and "mx" in Grade 11, which indicates that the Grades 10 and 11 estimates
are not constrained to be equal with each other. These parameter restrictions reduce
the number of estimated item-response probabilities from 48 to 40.

In Model 2 in Table 5.4, the item-response probabilities corresponding to the
questions about stealing are constrained to be equal, and measurement invariance

across the cohorts is specified. Now, to specify a Model 1 that is appropriate for the hypothesis test, the equivalence sets specified in Model 1 must also be constrained to be equal across groups in addition to the constraints on individual item-response probabilities. In other words, the item-response probabilities corresponding to the questions about stealing must be specified to be equal to each other both within cohorts and across cohorts. This is specified in Table 5.4 by means of identical letters (m, n, o, and p) within and across cohorts for these questions. With these parameter restrictions specified, a total of 20 item-response probabilities are estimated.

5.8.2   Test of measurement invariance in the delinquency example

Now let us return to the adolescent delinquency example. To test whether there is measurement invariance across the two cohorts, Grades 10 and 11, two different models were fit and compared. In Model 1, the item-response probabilities were allowed to vary across the groups. In Model 2, equivalence sets were set up in the manner illustrated in Table 5.3 so that the item-response probabilities were constrained to be equal across the groups. Table 5.5 shows the resulting likelihood-ratio statistic G^2, the AIC, and the BIC for the delinquency example, as well as the G^2 difference test. The G^2 difference test ($G^2_\Delta = 27.8, df = 24, p = .27$) is not significant, suggesting that the item-response probabilities are not significantly different across groups. In addition, the AIC and BIC both suggest that Model 2 provides a more optimal balance of model fit and parsimony. Therefore, we can comfortably conclude that measurement invariance holds across Grades 10 and 11. At this point it would be appropriate either to proceed with a multiple-group model with the item-response probabilities constrained to be equal across groups (if group differences in latent class prevalences are of interest), or to model both cohorts together without including cohort as a grouping variable.

Table 5.5   Fit Statistics for Test of Measurement Invariance for Adolescent Delinquency Example (Add Health Public-Use Data, Wave I; $N = 2,087$)

	G^2	df	AIC	BIC	ℓ
Model 1: Item-response probabilities vary across cohorts	81.1	73	189.1	493.9	−6,000.0
Model 2: Item-response probabilities equal across cohorts	108.9	97	168.9	338.2	−6,013.8
$G^2_2 - G^2_1 = 27.8, df = 24, p = .27$					

5.9 INTERPRETATION WHEN MEASUREMENT INVARIANCE DOES NOT HOLD

In the adolescent delinquency example above, we concluded that the item-response probabilities do not differ across groups, and then compared the prevalences of the latent classes across the groups. But what if there had been evidence of group differences in the item-response probabilities? Would it have made sense to examine group differences in the prevalences of the latent classes, and how should any observed differences have been interpreted? In this section and the next we discuss what to do when there is evidence of some degree of measurement invariance. Here we discuss allowing the item-response probabilities to vary and interpreting the latent classes accordingly. In Section 5.10 we discuss how to specify that measurement invariance holds for some variables or some subgroups.

If measurement invariance can be established, the latent classes have the same interpretation in all groups. Group comparisons involving latent variables are straightforward "apples-to-apples" comparisons and therefore are relatively easy to interpret. In contrast, whenever the item-response probabilities are different across groups in LCA, the question of how to interpret differences in the latent class prevalences arises. To the extent that measurement characteristics differ across groups, the latent classes may have different meanings across groups. In this case group comparisons may take on more of an "apples-to-oranges" flavor; in other words, there may be qualitative differences between the groups. Thus, to the extent that there are group differences in the item-response probabilities, any direct comparisons of the latent class prevalences must be done with any differences in the meaning of the latent classes in mind. To gain a sense of how to interpret the latent classes and their prevalences when the item-response probabilities differ between groups, consider the hypothetical set of item-response probabilities in Tables 5.6 and 5.7. Differences between the groups are in bold font.

Table 5.6 illustrates a situation in which there are clear and pronounced group differences in measurement. In the Non-/Mild Delinquents latent class, those in Grade 11 are much more likely to have lied to their parents than those in Grade 10. In the Verbal Antagonists latent class, those in Grade 11 were very likely to have taken part in a group fight, whereas those in Grade 10 were unlikely to have done so. In the Shoplifters latent class, those in Grade 10 were likely to have reported being publicly loud/rowdy/unruly, but those in Grade 11 were unlikely to report this behavior. In the General Delinquents latent class, tenth graders are slightly less likely to have taken part in a group fight.

The differences across groups in item-response probabilities shown in Table 5.6 suggest that a different interpretation of the latent classes is needed, one that allows the interpretation to differ across grades. For the first latent class the label Non-/Mild Delinquents fits well for Grade 10, but perhaps the label Liars Only would be more

Table 5.6   Hypothetical Item-Response Probabilities for Delinquency Example
Illustrating Pronounced Group Differences in Measurement

	Latent Class			
	Non-/Mild Delinquents	Verbal Antagonists	Shoplifters	General Delinquents
Grade 10				
Response = Yes				
Lied to parents	**.33***	.81	.78	.89
Publicly loud/rowdy/unruly	.20	.82	**.62**	1.00
Damaged property	.02	.25	.25	.89
Stolen something from store	.02	.02	.92	.88
Stolen something worth < $50	.00	.03	.73	.88
Taken part in group fight	.04	**.31**	.24	**.64**
Grade 11				
Response = Yes				
Lied to parents	**.80**	.81	.78	.89
Publicly loud/rowdy/unruly	.20	.82	**.20**	1.00
Damaged property	.02	.25	.25	.89
Stolen something from store	.03	.02	.92	.88
Stolen something worth < $50	.00	.03	.73	.88
Taken part in group fight	.04	**.85**	.24	**.78**

* Item-response probabilities in bold differ between cohorts.

suitable for Grade 11. For the second latent class, those in Grade 11 do not seem to be merely verbal antagonists; a label like Rowdy Behavior might be more descriptive. Similarly, the Shoplifters latent class could be labeled Rowdy Shoplifters in Grade 10 and Quiet Shoplifters in Grade 11. Thus each grade has its own array of latent classes and interpretations. There is some overlap between the two arrays—the General Delinquents latent class is essentially the same across the two grades—but the other latent classes are different.

Because of these differences, although it is appropriate to note the prevalences associated with each array, it probably does not make sense to make a direct comparison between, say, the prevalence associated with the first latent class for Grade 10 and the corresponding prevalence for Grade 11. This would be the kind of "apples-to-oranges" comparison mentioned above, in other words, the comparison involves not only quantitative differences in prevalence but qualitative differences in the meaning of the latent class. However, a comparison across cohorts of the prevalence associated with the General Delinquents latent class may be meaningful here, even in the presence of measurement differences in the other three latent classes.

The presence of group differences in item-response probabilities does not always rule out comparisons of latent class prevalences, but it does mean that the comparisons must be made cautiously. For example, Table 5.7 illustrates somewhat more modest group differences (the question of whether statistically significant group differences in item-response probabilities should be considered meaningful is discussed in Section 5.11). Eleventh graders in the Non-/Mild Delinquents latent class are slightly more

Table 5.7   Hypothetical Item-Response Probabilities for Delinquency Example Illustrating Moderate Group Differences in Measurement

	Latent Class			
	Non-/Mild Delinquents	Verbal Antagonists	Shoplifters	General Delinquents
Grade 10				
Response = Yes				
Lied to parents	**.33***	.81	.78	.89
Publicly loud/rowdy/unruly	.20	.82	**.62**	1.00
Damaged property	.02	.25	.25	.89
Stolen something from store	.03	.02	.92	.88
Stolen something worth < \$50	.00	.03	.73	.88
Taken part in group fight	.04	**.31**	.24	**.64**
Grade 11				
Response = Yes				
Lied to parents	**.45**	.81	.78	.89
Publicly loud/rowdy/unruly	.20	.82	**.57**	1.00
Damaged property	.02	.25	.25	.89
Stolen something from store	.03	.02	.92	.88
Stolen something worth < \$50	.00	.03	.73	.88
Taken part in group fight	.04	**.48**	.24	**.78**

*Item-response probabilities in bold differ between cohorts.

likely to have lied to their parents than are tenth graders in this latent class. In addition, in both the Verbal Antagonists and General Delinquents latent classes, those in Grade 11 are more likely to have taken part in a group fight than those in Grade 10, and Shoplifters in Grade 11 are a little less likely to have been publicly loud/rowdy/unruly.

Although the item-response probabilities in Table 5.7 are different across the two cohorts, overall the general pattern formed is about the same across the two groups. Thus the interpretation of the latent classes is consistent across groups. Whether the latent classes can be considered similar enough across groups to justify direct comparisons of latent class prevalences is a judgment that only the investigators can make. If the pattern is considered essentially identical, then cautious comparisons of the latent class prevalences may be made to assess group differences in latent class prevalences. However, it is important to keep in mind that the prevalences being compared correspond to latent classes that are very similar but not identical. For example, in noting that there are more Non-/Mild Delinquents in Grade 11 as compared to Grade 10, it is also important to note that the Non-/Mild Delinquents in Grade 11 are more likely to report having lied to their parents about their whereabouts and companions than are those in this latent class in Grade 10.

5.10   STRATEGIES WHEN MEASUREMENT INVARIANCE DOES NOT HOLD

5.10.1   Partial measurement invariance

Rejection of the null hypothesis of measurement invariance implies that anywhere from one to all of the item-response probabilities differ among groups. If most of the item-response probabilities are identical with only a few varying, it may be possible to establish that there is partial measurement invariance. In this case it may be helpful to constrain the item-response probabilities that can reasonably be considered identical and let the remaining item-response probabilities vary across groups. This will help to minimize the interpretational differences between the groups, although of course if the unconstrained item-response probabilities vary widely across the groups there may still be conceptually important group differences in the nature of the latent classes. Table 5.8 illustrates a set of parameter restrictions for the adolescent delinquency example in which only four item-response probabilities are allowed to vary between cohorts and the rest are constrained to be equal across cohorts. This approach might yield a solution similar to that shown in Table 5.7. When some of the item-response probabilities are constrained to be equal across groups, fewer parameters are estimated than in LCA in which all of the item-response probabilities vary across groups, but more than in LCA in which complete measurement invariance can be assumed. (For a discussion of partial measurement invariance in factor analysis, see Millsap and Kwok, 2004.)

In some cases, the item-response probabilities may be identical across groups except for those corresponding to a subset consisting of one or more observed variables. Suppose in the adolescent delinquency example it appeared that measurement invariance held for all of the variables except Damaged Property. Table 5.9 illustrates parameter restrictions in which the item-response probabilities for the Damaged Property variable may vary across grades, and all the other item-response probabilities are identical across grades. This model may be compared to one in which all the item-response probabilities are free to vary across groups, in order to test the hypothesis that the more parsimonious restricted model fits as well as the less restricted model. Alternatively, this model may be compared to one in which all the item-response probabilities are restricted to be equal across groups (i.e., a model with the parameter restrictions shown in Table 5.3). This would test the hypothesis that the more parsimonious completely restricted model in Table 5.3 fits as well as the less restricted model that allows the item-response probabilities for the Damaged property variable to vary across groups illustrated in Table 5.9.

Table 5.8   Parameter Restrictions Constraining Most, But Not All, Item-Response Probabilities to Be Equal Across Cohorts in Adolescent Delinquency Example

	Latent Class			
	Non-/Mild Delinquents	Verbal Antagonists	Shoplifters	General Delinquents
Grade 10				
Response = Yes				
Lied to parents	*	b	c	d
Publicly loud/rowdy/unruly	e	f	*	h
Damaged property	i	j	k	l
Stolen something from store	m	n	o	p
Stolen something worth < $50	q	r	s	t
Taken part in group fight	u	*	w	*
Grade 11				
Response = Yes				
Lied to parents	*	b	c	d
Publicly loud/rowdy/unruly	e	f	*	h
Damaged property	i	j	k	l
Stolen something from store	m	n	o	p
Stolen something worth < $50	q	r	s	t
Taken part in group fight	u	*	w	*

Note. Item-response probabilities designated with the same letter form an equivalence set.

* denotes a free parameter.

Table 5.9   Parameter Restrictions Constraining Item-Response Probabilities to Be Equal Across Cohorts for a Subset of Variables in Adolescent Delinquency Example

	Latent Class			
	Non-/Mild Delinquents	Verbal Antagonists	Shoplifters	General Delinquents
Grade 10				
Response = Yes				
Lied to parents	a	b	c	d
Publicly loud/rowdy/unruly	e	f	g	h
Damaged property	*	*	*	*
Stolen something from store	m	n	o	p
Stolen something worth < $50	q	r	s	t
Taken part in group fight	u	v	w	x
Grade 11				
Response = Yes				
Lied to parents	a	b	c	d
Publicly loud/rowdy/unruly	e	f	g	h
Damaged property	*	*	*	*
Stolen something from store	m	n	o	p
Stolen something worth < $50	q	r	s	t
Taken part in group fight	u	v	w	x

Note. Item-response probabilities designated with the same letter form an equivalence set.

* denotes a free parameter.

5.10.2   When measurement invariance holds
in a subset of groups

In situations in which there are more than two groups, measurement invariance may hold for some groups and not others. It is possible to test the hypothesis that measurement invariance holds across a subset of groups, which for clarity we call the equivalence subset, but that any remaining groups have item-response probabilities that differ from those in the equivalence subset.

This hypothesis can be tested by specifying a model in which the item-response probabilities for the equivalence subset are placed in appropriate equivalence sets, and the item-response probabilities for any remaining groups not in the equivalence subset are allowed to vary. For example, in the adolescent delinquency example, suppose that there was a Grade 12 cohort in addition to the Grades 10 and 11 cohorts, for a total of three cohorts. Further suppose that a test of measurement invariance that compares Model 1, in which all item-response probabilities are freely estimated, against Model 2, in which the item-response probabilities are constrained to be equal across all three cohorts, suggests that measurement invariance does not hold. Examination of the freely estimated item-response probabilities shows that the parameters are similar in Grades 10 and 11 but different in Grade 12.

The hypothesis that the item-response probabilities are identical in Grades 10 and 11 but different in Grade 12 can be tested as follows: A new model, Model 3, is created. Model 3 is identical to Model 2 except that the Grade 12 item-response probabilities are freely estimated rather than placed in an equivalence set along with those belonging to the Grades 10 and 11 cohorts. A specification of the appropriate equivalence sets for Model 3 is illustrated in Table 5.10. The fit of Model 3 can then be compared to the fit of Model 1 using the difference G^2 test. If this test is not significant, the item-response probabilities can be considered identical in Grades 10 and 11 but different in the Grade 12 cohort.

5.11   SIGNIFICANT DIFFERENCES AND
IMPORTANT DIFFERENCES

As shown above, the analytic steps that must be taken to test statistically whether measurement invariance holds are straightforward. However, the judgment of whether measurement invariance holds sufficiently to make direct comparisons between groups requires considerable thought and often cannot be made solely on the basis of statistical criteria. This does not mean that statistical tests are unnecessary. Rather, it means that statistical tests are an important decision-making tool, but only part of the picture. Group differences may be statistically significant, but conceptually unimportant; conversely, they may be nonsignificant, but in the investigator's judgment potentially important.

Table 5.10   Parameter Restrictions Constraining Item-Response Probabilities to Be Equal Across Only Grades 10 and 11 in a Hypothetical Adolescent Delinquency Example with Three Grades

	Latent Class			
	Non-/Mild Delinquents	Verbal Antagonists	Shoplifters	General Delinquents
Grade 10				
Response = Yes				
Lied to parents	a	b	c	d
Publicly loud/rowdy/unruly	e	f	g	h
Damaged property	i	j	k	l
Stolen something from store	m	n	o	p
Stolen something worth < $50	q	r	s	t
Taken part in group fight	u	v	w	x
Grade 11				
Response = Yes				
Lied to parents	a	b	c	d
Publicly loud/rowdy/unruly	e	f	g	h
Damaged property	i	j	k	l
Stolen something from store	m	n	o	p
Stolen something worth < $50	q	r	s	t
Taken part in group fight	u	v	w	x
Grade 12				
Response = Yes				
Lied to parents	*	*	*	*
Publicly loud/rowdy/unruly	*	*	*	*
Damaged property	*	*	*	*
Stolen something from store	*	*	*	*
Stolen something worth < $50	*	*	*	*
Taken part in group fight	*	*	*	*

Note. Item-response probabilities designated with the same letter form an equivalence set.

∗ denotes a free parameter.

A key consideration in deciding whether significant differences are important is that the power of the G^2 difference test varies considerably depending on the sample size. Even small group differences in estimates of item-response probabilities may result in a statistically significant G^2 difference test if the sample size is large. Thus it is important to examine the magnitude and overall pattern of any differences, irrespective of the outcome of any statistical tests. If the overall pattern of differences suggests that the interpretation of the latent classes is essentially identical and the differences are not large in magnitude, the investigator may wish to conclude that although the differences may be statistically significant, they are not important. In this case, it may make sense to impose measurement invariance.

Most investigators who use LCA wish to constrain item-response probabilities across groups whenever it is reasonable to do so. This is because multiple-group latent class models in which measurement invariance is specified are more parsimonious, and tend to be easier to interpret and better identified, than corresponding models in which the item-response probabilities are free to vary, and therefore estimation of many more parameters is required. All of these characteristics are desirable, particularly when the research agenda calls for adding covariates to the model, as discussed later in the book. Thus sometimes when the hypothesis test for measurement invariance is significant but the observed differences are sufficiently small in magnitude so as not to have a major impact on the interpretation of the latent classes, the investigator might choose to "overrule" the hypothesis test and constrain the item-response probabilities to be equal across groups. An example of this is provided in Section 5.11.1.

Whether differences that are statistically meaningful are conceptually important, and whether one model is more appealing than another on theoretical or substantive grounds, are judgment calls that must be made by the investigator. In general, there is nearly always a trade-off between parsimony, ease of interpretation, and stability on the one hand and more accurate model specification on the other. Although some of the details may be unique to the latent class field, this trade-off is a familiar one in many areas of statistical modeling of behavioral data.

5.11.1   Empirical example:
Positive health behaviors

In this section we present an example of a situation in which analysis of empirical data presented a difficult decision about whether or not to conclude that measurement could appropriately be constrained to be equal across groups. We realize that readers may or may not agree with the decisions we made! Nevertheless, we want to share the thought process in the hope that it will be informative.

Recall from Chapter 4 the latent class model of positive health behaviors in $N = 2,065$ high school seniors in the Monitoring the Future 2004 12th-grade cohort (Johnston et al., 2005). The variables in the LCA are five indicators of health-related behaviors: Eats breakfast; Eats at least some green vegetables; Eats at least some fruit; Exercises vigorously; and Gets at least seven hours sleep. Each indicator was coded 1 for "never/seldom engage in behavior," 2 for "sometimes/most days engage in behavior," and 3 for "nearly every day or every day engages in behavior." Based on the procedure reported in Chapter 4, a model with five latent classes was selected. Table 5.11 shows the latent class prevalences and item-response probabilities for this model. Individuals in the Unhealthy latent class, which has a prevalence of 14 percent, were most likely to respond that they never or seldom eat breakfast, eat green vegetables, eat fruit, exercise vigorously, or get at least seven hours of sleep. Those in the Sleep Deprived latent class, with a prevalence of 12 percent, were similar to the Unhealthy latent class, except that they were more likely to report eating vegetables and fruits. About 37 percent of the sample was in the Typical latent class, which was characterized by engaging in each of the positive health behaviors sometimes or most days. Healthy Eaters, constituting about 11 percent of the sample, were similar to the Typical latent class with respect to eating breakfast and exercising. They were

Table 5.11   Five Latent Classes of Positive Health Behaviors (Monitoring the Future Data, 2004; $N = 2,065$)

	Latent Class				
	Unhealthy	Sleep Deprived	Typical	Healthy Eaters	Healthy
Latent class prevalences	.14	.12	.37	.11	.27
Item-response probabilities					
Eats breakfast					
Never/seldom	**.78**[*]	**.59**	.32	.39	.10
Sometimes/most days	.19	.29	.46	.41	.17
Nearly every day/every day	.04	.12	.23	.21	**.73**
Eats at least some green vegetables					
Never/seldom	**.90**	.00	.12	.00	.02
Sometimes/most days	.09	**1.00**	**.79**	.28	.14
Nearly every day/every day	.01	.00	.09	**.71**	**.84**
Eats at least some fruit					
Never/seldom	**.65**	.10	.06	.01	.01
Sometimes/most days	.32	**.90**	**.86**	.23	.09
Nearly every day/every day	.03	.00	.09	**.76**	**.91**
Exercises vigorously					
Never/seldom	**.59**	**.56**	.16	.19	.07
Sometimes/most days	.30	.38	**.53**	**.54**	.31
Nearly every day/every day	.11	.06	.31	.27	**.62**
Gets at least seven hours sleep					
Never/seldom	**.55**	**.70**	.09	**.60**	.08
Sometimes/most days	.36	.28	**.63**	.36	.40
Nearly every day/every day	.08	.02	.28	.05	**.52**

$\ell = -9,533.1$.

[*] Item-response probabilities > .5 in bold to facilitate interpretation.

more likely than the Typical latent class to eat vegetables and fruit, but less likely to report getting at least seven hours of sleep. The Healthy latent class had a prevalence of 27 percent. Individuals in this latent class were likely to report engaging in each of the healthy behaviors nearly every day or every day.

Suppose that we are now interested in examining whether there are gender differences in profiles of health behaviors among high school seniors. Further suppose it has been established that a five-latent-class solution is appropriate for both males and females. Now we wish to establish whether or not it is reasonable to impose measurement invariance across genders. This was done by comparing two models, Models 1 and 2, using the approach described above. In Model 1, all item-response probabilities were allowed to vary across gender, and in Model 2 the item-response probabilities were constrained to be equal across genders using the approach depicted in Table 5.3.

Table 5.12 shows the fit statistics for these two models. The difference G^2 was statistically significant ($G^2_\Delta = 85.0, df = 50, p < .002$), suggesting that the null hypothesis that measurement invariance held across gender should be rejected.

Table 5.12   Fit Statistics for Test of Measurement Invariance Across Genders for Latent Class Model of Positive Health Behaviors (Monitoring the Future Data, 2004; $N = 2,065$)

	G^2	df	AIC	BIC	ℓ
Model 1: Item-response probabilities free to vary across genders	448.8	377	664.8	1,273.1	−9,459.3
Model 2: Item-response probabilities constrained equal across genders	533.8	427	649.8	976.5	−9,501.8
$G_2^2 - G_1^2 = 85.0, df = 50, p < .002$					

However, both the AIC and BIC pointed toward the more parsimonious Model 2. Thus the model fit indices suggested different decisions about measurement invariance.

As discussed above, it is always a good idea to examine the estimates of the item-response probabilities in each group in order to gain a sense of how similar or different they are, and whether the overall pattern of estimates, and therefore interpretation of the latent classes, appears at least fairly similar. This becomes particularly important when the likelihood-ratio difference test G^2, AIC, and BIC do not all agree, as was the case in the analyses reported here. (In our experience with fitting multiple-group latent class models in empirical data, we have found that the AIC and BIC may be more helpful than the difference G^2, particularly if the difference G^2 is associated with large df.)

The latent class prevalences for Model 1 appear in Table 5.13, and the item-response probabilities for Model 1 appear in Table 5.14. In general, the overall pattern in the item-response probabilities appeared to be consistent across genders, and observed gender differences were minimal. Most important, the pattern of item-response probabilities suggested identical latent class labels for males and females. For example, the label Unhealthy was appropriate for the first latent class for both males and females. Both genders in this latent class were most likely to respond "Never/seldom" to all five positive health behaviors. The item-response probabilities were also very similar for the Sleep Deprived and Typical latent classes. The fourth and fifth latent classes each included some relatively minor gender differences. In

Table 5.13   Latent Class Prevalences for Model of Positive Health Behaviors with Item-Response Probabilities Allowed to Vary Across Genders (Monitoring the Future Data, 2004; $N = 2,065$)

	Latent Class				
	Unhealthy	Sleep Deprived	Typical	Healthy Eaters	Healthy
Males	.10	.09	.43	.17	.22
Females	.17	.14	.29	.14	.26

Table 5.14   Item-Response Probabilities Allowed to Vary Across Genders for Model of Positive Health Behaviors (Monitoring the Future Data, 2004; $N = 2,065$)

	Latent Class				
	Unhealthy	Sleep Deprived	Typical	Healthy Eaters	Healthy
			Males		
Eats breakfast					
Never/seldom	**.77**[*]	**.82**	.35	.34	.05
Sometimes/most days	.21	.07	.43	**.56**	.00
Nearly every day/every day	.02	.11	.22	.11	**.95**
Eats at least some green vegetables					
Never/seldom	**.96**	.31	.14	.10	.00
Sometimes/most days	.04	**.62**	**.78**	.29	.15
Nearly every day/every day	.00	.06	.08	**.61**	**.85**
Eats at least some fruit					
Never/seldom	**1.00**	.05	.05	.04	.00
Sometimes/most days	.00	**.89**	**.93**	.12	.09
Nearly every day/every day	.00	.07	.02	**.84**	**.91**
Exercises vigorously					
Never/seldom	**.69**	**.60**	.14	.06	.02
Sometimes/most days	.19	.40	**.53**	.29	.32
Nearly every day/every day	.12	.00	.33	**.65**	**.66**
Gets at least seven hours sleep					
Never/seldom	**.55**	**.79**	.16	.24	.08
Sometimes/most days	.36	.21	**.61**	.44	.35
Nearly every day/every day	.09	.00	.23	.32	**.57**
			Females		
Eats breakfast					
Never/seldom	**.73**	**.55**	.29	.47	.06
Sometimes/most days	.21	.32	.48	**.52**	.14
Nearly every day/every day	.06	.13	.23	.01	**.81**
Eats at least some green vegetables					
Never/seldom	**.75**	.00	.08	.00	.02
Sometimes/most days	.23	**1.00**	**.84**	.34	.14
Nearly every day/every day	.02	.00	.08	**.66**	**.85**
Eats at least some fruit					
Never/seldom	**.56**	.07	.06	.00	.01
Sometimes/most days	.41	**.92**	**.87**	.24	.10
Nearly every day/every day	.03	.00	.08	**.76**	**.89**
Exercises vigorously					
Never/seldom	**.55**	**.55**	.22	.22	.13
Sometimes/most days	.35	.37	**.56**	**.54**	.36
Nearly every day/every day	.10	.09	.22	.25	**.51**
Gets at least seven hours sleep					
Never/seldom	**.51**	**.82**	.00	.49	.14
Sometimes/most days	.40	.18	**.69**	.35	.44
Nearly every day/every day	.09	.00	.31	.16	.42

$\ell = -9,459.3$.

[*]Item-response probabilities >.5 in bold to facilitate interpretation.

the fourth latent class the modal response to the question about exercising vigorously was "Nearly every day" for males and "Sometimes/most days" for females. Another difference occurred on the question asking about whether the individual gets at least seven hours of sleep. The modal response for males was "Sometimes/most days," whereas for females it was "Never/seldom." Despite these differences, the label Healthy Eaters still seemed appropriate for both genders. In the fifth latent class, both genders were likely to respond "Nearly every day/every day" to all of the questions, with one exception: Females in this latent class were more likely to choose the response "Sometimes/most days" than they were to choose either of the other response alternatives to the question about whether they get at least seven hours of sleep. Again, despite this difference, the latent class label Healthy still seemed appropriate.

Based on the similarities in interpretation, as well as the lower AIC and BIC values in Model 2, we opted to impose measurement invariance across genders before making gender comparisons in the prevalence of healthy behavior latent classes. Table 5.15 shows the LCA parameter estimates when the item-response probabilities were constrained to be equal across groups.

Table 5.15   Latent Class Model of Positive Health Behaviors with Item-Response Probabilities Constrained Equal Across Genders (Monitoring the Future Data, 2004; $N = 2,065$)

	Latent Class				
	Unhealthy	Sleep Deprived	Typical	Healthy Eaters	Healthy
Latent class prevalences					
Males	.12	.11	.42	.08	.27
Females	.08	.24	.28	.24	.15
Item-response probabilities					
Eats breakfast					
Never/seldom	**.78***	**.61**	.32	.26	.10
Sometimes/most days	.19	.28	.46	.35	.14
Nearly every day/every day	.03	.11	.22	.38	**.75**
Eats at least some green vegetables					
Never/seldom	**1.00**	.15	.12	.00	.03
Sometimes/most days	.00	**.83**	**.79**	.23	.14
Nearly every day/every day	.00	.02	.09	**.77**	**.84**
Eats at least some fruit					
Never/seldom	**.77**	.15	.05	.02	.00
Sometimes/most days	.19	**.84**	**.85**	.18	.07
Nearly every day/every day	.03	.01	.10	**.80**	**.92**
Exercises vigorously					
Never/seldom	**.60**	**.61**	.10	.25	.01
Sometimes/most days	.28	.36	**.55**	**.52**	.25
Nearly every day/every day	.13	.03	.35	.23	**.74**
Gets at least seven hours sleep					
Never/seldom	**.55**	**.54**	.14	.37	.07
Sometimes/most days	.36	.38	**.60**	.41	.39
Nearly every day/every day	.09	.07	.26	.22	**.54**

$\ell = -9,501.8$.

*Item-response probabilities >.5 in bold to facilitate interpretation.

5.11.1.1 Conclusions about gender differences in health behaviors

The original research question was whether there are gender differences in profiles of health behaviors among high school seniors. As we have discussed, this question may refer to two different aspects of LCA: the item-response probabilities or the latent class prevalences. By settling on the model shown in Table 5.15, we concluded that there were not significant gender differences in the item-response probabilities. This conclusion implies that any differences between genders in health behaviors were quantitative and could be expressed in terms of differences in latent class prevalences. As Table 5.15 shows, the observed differences in latent class prevalences were large. Both genders were most likely to be in the Typical latent class, but over 40 percent of males were in this latent class as opposed to about 28 percent of females. For females the next most prevalent latent classes were Sleep Deprived (24%) and Healthy Eaters (24%), whereas for males these were the least prevalent. Not surprisingly, Table 5.16 shows that a test for overall differences in latent class prevalences is highly significant ($G_\Delta^2 = 63.4, df = 4, p < .0001$). The latent class prevalences for each gender are illustrated in Figure 5.1.

Our objective in presenting this somewhat controversial example is to give the reader a sense of the kind of decision making that is frequently required when multiple-group LCA is applied to empirical data in the social, behavioral, and health sciences. Did we make the right decision by concluding that measurement invariance can be assumed across genders? On the one hand, this decision simplifies the model and makes comparing latent class prevalences across genders straightforward. It also nicely sets the stage for our introduction of covariates to help explain gender differences in latent class prevalences in Chapter 6. On the other hand, constraining the item-response probabilities to be equal across genders risks oversimplifying the model and thereby masking important and fundamental gender differences in the very definitions of the latent classes. Our view is that in this example, the simpler and more parsimonious model is a reasonable representation of the data, and there is

Table 5.16 Fit Statistics for Test of Gender Differences in Latent Class Prevalences for Latent Class Model of Positive Health Behaviors (Monitoring the Future Data, 2004; $N = 2,065$)

	G^2	df	AIC	BIC	ℓ
Model 1: Latent class prevalences free to vary across genders	533.8	427	649.8	976.5	−9,501.8
Model 2: Latent class prevalences constrained equal across genders	597.2	431	705.2	1,009.4	−9,533.5
$G_2^2 - G_1^2 = 63.4, df = 4, p < .0001$					

Note. Item-response probabilities constrained equal across genders.

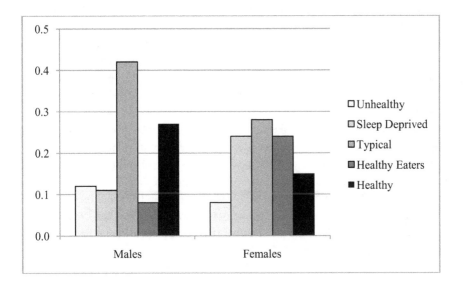

Figure 5.1   Prevalence of latent classes of positive health behaviors by gender (Monitoring the Future data, 2004; $N = 2,065$).

only a small risk of model misspecification serious enough to lead to an erroneous scientific conclusion.   However, we also acknowledge that in the course of this exercise we made several judgment calls with which the reader may take exception. Other approaches are possible; an intermediate approach would have been to impose partial measurement invariance, constraining all item-response probabilities to be equal across genders except the ones noted as exceptions above, that is, the item-response probabilities corresponding to the question about sleep for the Healthy Eaters and Healthy latent classes, and those corresponding to the question about exercise for the Healthy Eaters latent class. Whether or not the reader agrees with our decisions, we hope that seeing the decision-making process has been illuminating.

5.12   TESTING EQUIVALENCE OF LATENT CLASS PREVALENCES ACROSS GROUPS

Another question that is often of interest is whether the prevalences of latent classes are the same or different across groups. The equivalence of latent class prevalences, that is, the equivalence of the γ parameters across groups, can be examined using an approach very similar to the one that was used above to test for measurement invariance across groups. Although similarities or differences in latent class prevalences across groups may be worth general consideration under any circumstances, a hypothesis test that involves a direct comparison of latent class prevalences across groups is easiest to interpret when, as in the adolescent delinquency example, strict measurement invariance can be assumed. Therefore, in general we recommend conducting this hypothesis test only when the item-response probabilities are constrained to be equal across groups.

Assuming that measurement invariance has been established, the question of whether latent class prevalences are identical across groups can be addressed by comparing the fit of Model 1, in which the item-response probabilities are constrained to be equal across groups and the latent class prevalences are free to vary, against the fit of Model 2, in which both the item-response probabilities and the latent class prevalences are constrained to be equal across groups. The hypothesis that Model 2 fits as well as Model 1 (i.e., that the class prevalences are not different across groups) can be tested via the difference G^2.

5.12.1 Empirical example: Adolescent delinquency

The first empirical example will involve a hypothesis test to assess whether the latent class prevalences in the adolescent delinquency example are equal across cohorts. Table 5.17 shows the latent class prevalences for the Grades 10 and 11 cohorts.

The hypothesis test involved comparing two models. Model 1 constrained the estimation of the item-response probabilities as shown in Table 5.3, and freely estimated latent class prevalences. Model 2 included the same constraints as Model 1 on the estimates of item-response probabilities. It differed from Model 1 because in addition it included constraints on the estimates of latent class prevalences like those in Table 5.18. These constraints restricted the estimates of the latent class prevalences to be equal across the two cohorts, resulting in a model with three fewer latent class prevalences being estimated.

Table 5.17 Latent Class Prevalences Across Cohorts in Four-Latent-Class Model of Past-Year Delinquency (Add Health Public-Use Data, Wave I; $N = 2,087$)

| | Latent Class | | | |
	Non-/Mild Delinquents	Verbal Antagonists	Shoplifters	General Delinquents
Grade 10	.47	.27	.19	.07
Grade 11	.51	.26	.18	.05

Table 5.18   Parameter Restrictions Constraining Latent Class Prevalences to Be Equal Across Cohorts in the Adolescent Delinquency Example (Add Health Public-Use Data, Wave I; N = 2,087)

	Latent Class			
	Non-/Mild Delinquents	Verbal Antagonists	Shoplifters	General Delinquents
Grade 10	a	b	c	d
Grade 11	a	b	c	d

Note. Item-response probabilities designated with the same letter form an equivalence set.

Table 5.19   Fit Statistics for Test of Cohort Differences in Latent Class Prevalences for Adolescent Delinquency Example (Add Health Public-Use Data, Wave I; N = 2,087)

	G^2	df	AIC	BIC	ℓ
Model 1: Latent class prevalences vary across cohorts	108.9	97	168.9	338.2	−6,013.8
Model 2: Latent class prevalences equal across cohorts	112.8	100	166.8	319.2	−6,015.8
$G_2^2 - G_1^2 = 3.9, df = 3, p = .27$					

Note. Item-response probabilities constrained equal across cohorts.

The results of the hypothesis test are shown in Table 5.19. As the table shows, the latent class prevalences did not vary significantly across cohorts ($G_\Delta^2 = 3.9, df = 3, p = .27$).

5.12.2 Empirical example: Health risk behaviors

The second empirical example revisits the health risk behaviors data first presented in Chapter 2. The example is based on data gathered from 13,840 U.S. high school students who participated in the Youth Risk Behavior Survey (Centers for Disease Control and Prevention, 2004) in 2005. Participants were asked to indicate whether they had engaged in each of 12 health risk behaviors. The LCA presented in Chapter 2 examined the sample as a whole. This LCA suggested that there were five latent classes: Low risk; Early experimenters; Binge drinkers; Sexual risk-takers; and High risk.

The participants in this study were in either Grade 9, 10, 11, or 12 in school. It may be interesting to examine whether the latent status prevalences vary across grades. This can be accomplished using multiple-group LCA.

The 12 health risk behaviors and the marginal response proportions of a "Yes" response for the four grades are shown in Table 5.20. Several general trends are evident in these marginals. The overall proportion of adolescents responding "Yes"

Table 5.20   Proportion of Students in Each Cohort Reporting Each Health Risk
Behavior (Youth Risk Behavior Survey, 2005; $N = 13,840$)

Grade N	Proportion Responding Yes			
	9 3,332	10 3,470	11 3,529	12 3,509
Health risk behavior				
Smoked first cigarette before age 13	.18	.16	.14	.13
Smoked daily for 30 days	.09	.10	.13	.15
Has driven when drinking	.06	.07	.12	.17
Had first drink before age 13	.34	.27	.22	.20
$\geq$ 5 drinks in a row in past 30 days	.18	.24	.26	.31
Tried marijuana before age 13	.12	.09	.08	.07
Used cocaine in life	.06	.08	.09	.10
Sniffed glue in life	.14	.13	.11	.09
Used meth in life	.06	.06	.06	.06
Used ecstasy in life	.06	.06	.06	.07
Had sex before age 13	.10	.08	.07	.06
Had sex with four or more people	.11	.13	.19	.24

Note. Proportions are based on N responding to each question. Amount of missing data varied across questions and grades.

to several of the items increases in the older grades, suggesting an increase in risky behavior. For example, the proportion who report having had five or more drinks in a row in the past 30 days ranges from about .18 in Grade 9 to about .31 in Grade 12, representing an increase of about 72 percent. Because these are cross-sectional data, it is impossible to determine the extent to which this reflects age-related developmental changes in risky behavior as opposed to cohort differences or historical trends. (Chapters 7 and 8 introduce some approaches to analysis of longitudinal data.)

Table 5.20 shows some troubling age-related trends. The proportion reporting a first experience with smoking, alcohol use, marijuana use, and sex before age 13 is greater in earlier grades, suggesting that individuals are initiating these behaviors at younger ages. The proportion of individuals reporting first cigarette use before age 13 is about 38 percent higher in the Grade 9 group than it is in the Grade 12 group. Similarly, the proportion reporting first use of alcohol before age 13 is about 70 percent higher; the proportion reporting first use of marijuana before age 13 is about 71 percent higher, and the proportion reporting having had sex before age 13 is about 67 percent higher in Grade 9 than in Grade 12. Perhaps a multiple-group LCA can be helpful in illuminating this complex array of data.

5.12.2.1 *Results of the multiple-group LCA*

A multiple-group LCA was performed, specifying five latent classes and constraining the item-response probabilities to be equal across grades. Table 5.21 contains the latent class prevalences for each grade. These prevalences are illustrated in Figure 5.2.

Table 5.21   Latent Class Prevalences Across Grades in Five-Latent-Class Model of
Health Risk Behaviors (Youth Risk Behavior Survey, 2005; $N = 13,840$)

	Latent Class				
	Low Risk	Early Experimenters	Binge Drinkers	Sexual Risk-takers	High Risk
Latent class prevalences					
Grade 9	.71	.18	.01	.04	.06
Grade 10	.69	.12	.09	.04	.06
Grade 11	.64	.07	.17	.05	.06
Grade 12	.60	.04	.24	.06	.06

Low Risk is the most prevalent latent class for each grade, although the prevalence declines in each successively older grade. The prevalence of the Early Experimenters latent class also declines in older grades. The prevalence of the Sexual Risk-Takers latent class increases only slightly as a function of grade, from a low of about 4 percent in Grades 9 and 10 to a high of about 6 percent in Grade 12. The prevalence of the High Risk latent class stays roughly constant across grades at about 6 percent. The latent class that shows the largest increase across grades is Binge Drinkers. Only about 1 percent of those in Grade 9 belong to this latent class, but there is a steady increase associated with each older grade, with nearly 25 percent of 12th graders classified as Binge Drinkers. The Binge Drinkers latent class in Grade 11 is roughly double the size of what it is in Grade 10.

The pattern of latent class prevalences across grades is suggestive of change over time. Because these are not longitudinal data, the pattern cannot be interpreted as change over time. However, it is possible to speculate about how change might be taking place. If the declines in the Low Risk and Early Experimenters latent classes mean that individuals are moving out of these latent classes, into which latent classes would they be moving? The High Risk latent class does not show any increase in prevalence across grades, so it seems unlikely that many individuals would be moving into this latent class. By the same logic the data suggest that a few individuals, but not many, could be moving into the Sexual Risk-Takers latent class. This suggests that the bulk of any movement between latent classes would be from the declining Low Risk and Early Experimenters latent class into the increasing Binge Drinkers latent class. We stress that these are speculations, not conclusions. They await longitudinal data for confirmation or disconfirmation. Importantly, as Table 5.20 shows, there appear to be historic trends occurring in which students in the lower grades are engaging in health risk behaviors before age 13 at a higher rate. Thus, it is likely that the prevalences of the latent classes reflecting greater health risk would be larger when those in the earlier grades reach Grade 12 than the corresponding prevalences based on those in the current Grade 12.

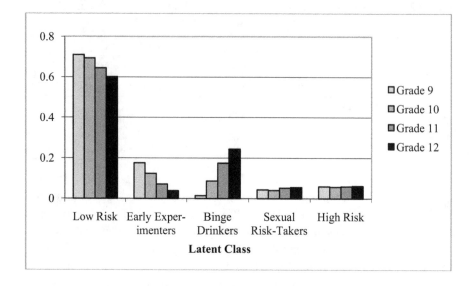

Figure 5.2   Prevalence of health risk behavior latent classes for each grade (Youth Risk Behavior Survey, 2005; $N = 13,840$).

5.12.2.2   Testing the hypothesis of equality of latent class prevalences across grades

The health risk behavior example can be used to demonstrate the use of the difference G^2 test to assess group differences in latent class prevalences. The model that was fit to produce the estimates of latent class prevalences that appear in Table 5.21 and Figure 5.2 did not include any parameter restrictions on the latent class prevalences (although, as mentioned above, it did constrain the item-response probabilities to be identical across grades); in other words, all latent class prevalences were freely estimated across grades. This model will be referred to as Model 1. To test for differences in latent class prevalences across grades, let us consider a set of models (Models 2 through 6), each of which includes one set of parameter restrictions equating prevalence of a single latent class across the four grades. Model 2 restricts the prevalence of the Low Risk latent class to be equal; Model 3 restricts the prevalence of Early Experimenters; Model 4 restricts the prevalence of Binge Drinkers; Model 5 restricts the prevalence of Sexual Risk-Takers; and Model 6 restricts the prevalence of the High Risk latent class. We then consider one final model (Model 7) in which latent class prevalence is constrained to be equal across grades for each of the five

Table 5.22   Freely Estimated and Restricted Latent Class Prevalences for Models of
Health Risk Behaviors (Youth Risk Behavior Survey, 2005; $N = 13,840$)

	Latent Class				
	Low Risk	Early Experi-menters	Binge Drinkers	Sexual Risk-Takers	High Risk
Model 1: Freely Estimated					
Grade 9	.71	.18	.01	.04	.06
Grade 10	.69	.12	.09	.04	.06
Grade 11	.64	.07	.17	.05	.06
Grade 12	.60	.04	.24	.06	.06
Model 2: Low Risk Constrained					
Grade 9	**.66**	.22	.01	.05	.07
Grade 10	**.66**	.15	.09	.05	.06
Grade 11	**.66**	.07	.16	.05	.06
Grade 12	**.66**	.03	.20	.05	.06
Model 3: Early Experimenters Constrained					
Grade 9	.77	**.11**	.02	.05	.06
Grade 10	.71	**.11**	.09	.04	.06
Grade 11	.64	**.11**	.16	.04	.06
Grade 12	.58	**.11**	.22	.04	.05
Model 4: Binge Drinkers Constrained					
Grade 9	.66	.14	**.12**	.04	.05
Grade 10	.66	.11	**.12**	.06	.05
Grade 11	.64	.09	**.12**	.10	.06
Grade 12	.61	.06	**.12**	.15	.07
Model 5: Sexual Risk-Takers Constrained					
Grade 9	.71	.17	.01	**.05**	.06
Grade 10	.69	.12	.09	**.05**	.06
Grade 11	.65	.07	.18	**.05**	.06
Grade 12	.60	.04	.25	**.05**	.06
Model 6: High Risk Constrained					
Grade 9	.71	.18	.01	.04	**.06**
Grade 10	.69	.12	.09	.04	**.06**
Grade 11	.64	.07	.17	.05	**.06**
Grade 12	.60	.04	.24	.06	**.06**
Model 7: All Latent Classes Constrained					
Grade 9	**.68**	**.09**	**.13**	**.04**	**.05**
Grade 10	**.68**	**.09**	**.13**	**.04**	**.05**
Grade 11	**.68**	**.09**	**.13**	**.04**	**.05**
Grade 12	**.68**	**.09**	**.13**	**.04**	**.05**

Note. For each model, bold numbers indicate latent class prevalences that were constrained to be equal
across grades.

latent classes. The latent class prevalences for each model are shown in Table 5.22
and the model fit statistics for each model appear in Table 5.23. Note that in Table
5.22, parameters that are part of an equivalence set appear in bold.

To understand the degrees of freedom associated with the G^2 difference tests that
are shown below, it is helpful to note how many latent class prevalences are estimated
in each model. In Model 1, in which there are no restrictions on the latent class
prevalences, there are four latent class prevalences estimated per grade, for a total

Table 5.23   Examination of the Impact of Parameter Restrictions on Model Fit for Model of Health Risk Behaviors (Youth Risk Behavior Survey, 2005; $N = 13{,}840$)

Latent Classes Constrained Equal Across Grades	G^2	df	AIC	BIC	ℓ
Model 1: None	7,517.9	16,307	7,669.9	8,242.6	−48,032.9
Model 2: Low Risk	7,579.2	16,310	7,725.2	8,275.3	−48,063.5
Model 3: Early Experimenters	7,694.5	16,310	7,840.5	8,390.6	−48,121.2
Model 4: Binge Drinkers	7,986.4	16,310	8,132.4	8,682.5	−48,267.1
Model 5: Sexual Risk-Takers	7,523.6	16,310	7,669.6	8,219.7	−48,035.7
Model 6: High Risk	7,518.3	16,310	7,664.3	8,214.4	−48,033.1
Model 7: All	8,143.4	16,319	8,271.4	8,753.7	−48,345.6

of 16. In Model 2, there is one equivalence set plus a number of freely estimated latent class prevalences. The equivalence set is comprised of the prevalences of the Low Risk latent class for each of the four grades. This equivalence set counts as one estimated parameter. Now let us turn to the freely estimated latent class prevalences. In the Grade 9 group, there are four latent class prevalences that are not part of the equivalence set. One of these can be obtained by subtraction, so of the remaining four latent class prevalences, three are estimated. By this reasoning there are three latent class prevalences estimated in each of the four grades. Thus the total number of latent class prevalences estimated in Model 2 is $1 + (3 \times 4) = 13$. Models 3 through 6 also each estimate 13 latent class prevalences.

In Model 7 each latent class prevalence is in one of five equivalence sets; in other words, the latent class prevalences are constrained to be equal across grades. Because one latent class prevalence can be obtained by subtraction, there are four latent class prevalences estimated in Model 7.

Each of Models 2 through 6, in which the prevalence of a single latent class is constrained to be equal across grades, is nested within the more general Model 1 (see Chapter 4 for a discussion of nesting). Therefore, a difference G^2 test can be used to compare Models 1 and 2, 1 and 3, 1 and 4, 1 and 5, and 1 and 6. In addition, the highly restricted Model 7 is nested within Model 1. (It is worth noting that Model 7 is also nested within Models 2 through 6.) However, note that among Models 2 through 6, no two models are statistically nested.

Table 5.24 shows the results of hypothesis tests comparing Model 1 to each other model. The G^2_Δ associated with each comparison of Model 1 to Models 2 through 6 has 3 degrees of freedom. As explained above, this is because whereas Model 1 estimates 16 latent class prevalences, Models 2 through 6 each estimate 13 latent class prevalences, or 3 fewer than Model 1. The G^2_Δ associated with the comparison of Model 1 to Model 7 has 12 degrees of freedom, because Model 7 estimates only four latent class prevalences, or 12 fewer than Model 1.

Based on the results in Table 5.24, we conclude that there is not sufficient evidence for differences across grades in the prevalence of the Sexual Risk-Takers and High

Table 5.24 Hypothesis Tests of Equality of Latent Class Prevalences Across Grades in Health Risk Behavior Example (Youth Risk Behavior Survey, 2005; $N = 13,840$)

Latent Class Prevalences Hypothesized Equal Across Grades	Comparison Model	Difference G^2	df	p-value	Conclusion: Reject H_0?
Low Risk	2	61.3	3	<.0001	Yes
Early Experimenters	3	176.6	3	<.0001	Yes
Binge Drinkers	4	468.5	3	<.0001	Yes
Sexual Risk-Takers	5	5.7	3	.13	No
High Risk	6	.4	3	.94	No
All five latent classes	7	625.5	12	<.0001	Yes

Note. Each model compared to Model 1.

Risk latent classes, but that we have very strong evidence for grade differences in the prevalence of the Low Risk, Early Experimenters, and Binge Drinkers latent classes. It is interesting to note that in this example both the AIC and BIC (see Table 5.23) agreed with the difference G^2 test. That is, both Models 5 and 6 had lower AIC and BIC values than Model 1, suggesting that they optimized the balance between parsimony and model fit better than Model 1. Models 2, 3, 4, and 7, on the other hand, had AIC and BIC values higher than those of Model 1, suggesting that the parameter restrictions imposed on those four models were not optimal.

5.13 SUGGESTED SUPPLEMENTAL READINGS

Clogg and Goodman (1985) and McCutcheon (2002) both provide additional information about multiple-group LCA. Lanza et al. (2007) includes information about how to fit multiple-group latent class models using Proc LCA.

5.14 POINTS TO REMEMBER

• Multiple-group LCA starts with the assumption that the sample of subjects at hand were drawn from two or more populations, that population membership is known, and that there may be scientifically interesting group differences in latent class prevalences or in both latent class prevalences and item-response probabilities.

• In multiple-group LCA estimates of latent class prevalences and item-response probabilities can be conditioned on group.

• Multiple-group LCA can be used to examine whether measurement invariance holds across groups. When measurement invariance can be assumed to hold across groups, it is appropriate to constrain the item-response probabilities to be equal across groups.

• All else being equal, models in which measurement invariance holds are more parsimonious than models in which measurement invariance does not hold.

• Before conducting multiple-group LCA it is a good idea to explore whether groups have the same number of latent classes. In making a decision about this, considerations include parsimony, statistical power, and the possibility of identification problems in one or more groups.

• Partial measurement invariance can be implemented by constraining some but not all item-response probabilities to be equal across groups.

• When there are more than two groups, it is possible to fit a model in which there is measurement invariance across a subset of the groups.

• Sometimes goodness-of-fit indices may disagree about whether it is appropriate to assume measurement invariance. In this case, parsimony, model identification, and ease of interpretation become particularly important considerations.

5.15   WHAT'S NEXT

In this chapter we discussed how to include a categorical grouping variable in LCA, in order to relate group membership to the item-response probabilities and the latent class prevalences. In the next chapter we discuss how to introduce covariates into LCA, both with and without a grouping variable.

CHAPTER 6

LATENT CLASS ANALYSIS
WITH COVARIATES

6.1 OVERVIEW

In this chapter we show how to introduce variables into LCA as covariates that predict latent class membership (Bandeen-Roche, Miglioretti, Zeger, and Rathouz, 1997 ; Dayton and Macready, 1988, 2002; van der Heijden, Dessens, and Böckenholt, 1996). This is accomplished by means of ordinary logistic regression; the only difference is that the outcome is latent rather than directly observed. We illustrate how covariates can be introduced using both multinomial and binomial logistic regression. The concept of odds and odds ratios are discussed, and we show how these relate to the intercepts and regression coefficients in logistic regression. Interactions between covariates will be fit, and we discuss how to interpret these interactions. We also compare and contrast LCA with covariates against multiple-group LCA, which was introduced in Chapter 5.

6.2   EMPIRICAL EXAMPLE: POSITIVE HEALTH BEHAVIORS

To begin thinking about how covariates can be used in LCA, let us return to the positive health behaviors empirical example from Chapters 4 and 5. This example is based on data collected from high school seniors (12th graders) as part of the 2004Monitoring the Future (Johnston et al., 2005) study. The variables in the LCA are five indicators of health-related behaviors: Eats breakfast; Eats at least some green vegetables; Eats at least some fruit; Exercises vigorously; and Gets at least seven hours sleep. Each indicator was coded 1 for "never/seldom engage in behavior," 2 for "sometimes/most days engage in behavior," and 3 for "nearly every day or every day engage in behavior." The baseline model as reported in Chapters 4 and 5 has five latent classes: Unhealthy, Sleep Deprived, Typical, Healthy Eaters, and Healthy.

In Chapter 5 we examined gender differences in this model. We concluded that it was reasonable to assume measurement invariance, and fit a model in which the item-response probabilities were constrained to be equal across groups but the latent class prevalences were allowed to vary. A hypothesis test indicated that there were significant gender differences in the latent class prevalences. The latent class prevalences and log-likelihoods for the baseline LCA model and the multiple-group LCA model originally reported in Chapter 5 appear in Table 6.1.

Let us now reframe the question about gender differences slightly. Instead of examining gender differences in a multiple-group framework, it is possible to introduce gender into the latent class model as a covariate. In this case males and females are not treated as separate populations, as they were in the multiple-group analysis in Chapter 5. Instead, males and females form a single group, and gender is used to predict latent class membership. A hypothesis test for gender differences in latent class prevalences can then be performed by testing whether the effect of the covariate is significant. The multiple-group approach and the covariate approach are compared and contrasted in Section 6.10.

Table 6.1   Latent Class Prevalences and Log-Likelihoods from Previously Fit Models of Positive Health Behaviors (Monitoring the Future Data, 2004; $N = 2{,}065$)

| | Latent Class Prevalences | | | | | |
	Unhealthy	Sleep Deprived	Typical	Healthy Eaters	Healthy	ℓ
Baseline model (Table 5.11)	.14	.12	.37	.11	.27	−9,533.1
Multiple-group model (Table 5.15)						−9,501.8
Males	.12	.11	.42	.08	.27	
Females	.08	.24	.28	.24	.15	

6.3 PREPARING TO CONDUCT LCA
WITH COVARIATES

The purpose of introducing covariates into a latent class model is to identify characteristics that predict membership in the various latent classes. In our view, before an investigator can attempt to introduce covariates, a necessary first step is to have a clear idea of what the latent structure is. For this reason, we recommend that before conducting LCA with covariates, it is a good idea to go through the procedures described earlier in this book to fit a baseline model without covariates. This model should provide an adequate representation of the data, have clearly interpretable latent classes, and be identified. Once this model has been arrived at, covariates may be introduced into the model.

When planning LCA with covariates, it can be helpful to think ahead about missing data. Some software for LCA with covariates allows missing data in the indicators of the latent classes but not in the covariates, and will automatically remove cases that have missing data on any covariate. If there are missing data on one or more covariates and steps are not taken to deal with this, the baseline model will be based on a larger, and therefore somewhat different, data set than the covariate model. There are two strategies for avoiding this problem. One strategy is to remove any cases with missing data on any of the covariates before fitting the baseline model, so that the same data set is used in all analyses. However, if this results in a severe loss of data, issues of bias and a loss in statistical power may arise. The other strategy is to use multiple imputation (Schafer, 1997; Schafer and Graham, 2002) to deal with missing data prior to any data analysis. This reduces or eliminates bias and preserves statistical power by enabling the investigator to make use of all available data. If multiple imputation is used, care must be taken to ensure that all covariates, along with any interactions between covariates, are present in the data set when the multiple imputation procedure is carried out. See Section 4.2.3 for more information about missing data and LCA.

6.3.1 Preparing variables for use
as covariates

Covariates are incorporated into LCA using a logistic regression framework (a thorough treatment of logistic regression may be found in Agresti, 1990). As in any regression framework, the set of covariates can include categorical variables, numeric variables (i.e., variables that are treated as continuous), or a combination of both.

Because all covariates are treated as numeric, categorical predictor variables must be "dummy coded" before they are entered into the regression. Dummy coding involves re-expressing the information in a categorical predictor variable as one or more binary (0/1) variables. The recoded binary variables are called *dummy variables*.

Table 6.2   Example Coding Scheme to Represent a Covariate with Three Response Categories

Political Affiliation	Dummy Variable 1	Dummy Variable 2
Democrat	1	0
Republican	0	1
Independent/Other	0	0

A categorical predictor with D categories is entered into a standard regression model as $D-1$ dummy variables (e.g., Cliff, 1987). Thus when a variable has two categories, such as gender, a single dummy-coded variable is required. In the health behavior example we chose to code females as a 0 and males as a 1.

If a covariate—for example, a question asking about political affiliation in the United States—had three response categories, a set of two dummy variables would be required in the model to represent the covariate. Table 6.2 shows one possible coding strategy that could be used to represent political affiliation using two dummy variables. In this coding strategy, Dummy Variable 1 takes on a value of 1 for Democrats and 0 for everyone else, and Dummy Variable 2 takes on a value of 1 for Republican and 0 for everyone else. Table 6.2 does not include a dummy variable for the Independent/Other group, which means that the Independent/Other group will serve as a comparison category. In this case, the regression coefficient for Dummy Variable 1 would be interpreted as the effect of being a Democrat compared to being Independent/Other, and the regression coefficient for Dummy Variable 2 would be interpreted as the effect of being a Republican compared to Independent/Other. The joint test of significance of the two dummy variables would reflect the relation between political affiliation and the dependent variable. An estimate of the effect of being Democrat as compared to Republican could be obtained using a slightly different coding scheme in which the Republican category is coded 0 in both dummy variables and the Independent/Other category is coded 0 in Dummy Variable 1 and 1 in Dummy Variable 2. Note that the choice of the comparison category is arbitrary and does not affect hypothesis testing, although sometimes this choice has an impact on how easily the results can be interpreted.

In general it is not necessary to to transform numeric variables before entering them as predictors in a logistic regression. However, we recommend standardizing numeric covariates, particularly if several covariates are to be included. When the predictors are standardized, a one-unit change translates to a one-standard-deviation change for each predictor variable. This makes it easier to compare effects across covariates.

6.4   LCA WITH COVARIATES: MODEL AND NOTATION

Before turning to the results of the empirical analysis, let us express the model and notation more formally.

As in previous chapters, suppose that there are $j = 1, ..., J$ observed variables measuring the latent classes, and that observed variable j has $r_j = 1, ..., R_j$ response categories. In the health behavior example there are $J = 5$ observed variables, and $R_j = 3$ for all j. Let $\mathbf{y} = (r_1, ..., r_j)$ represent the vector of a particular subject's responses to the J variables. Let L represent the latent variable with $c = 1, ..., C$ latent classes. As discussed above, in the present example $C = 5$. In addition, there is a covariate X, which in the example is gender. This covariate is to be used to predict latent class membership. Finally, $I(y_j = r_j)$ is an indicator function that equals 1 when the response to variable $j = r_j$, and equals 0 otherwise. (As mentioned in previous chapters, this function is merely a device for picking out the appropriate ρ parameters to multiply together.) Then the latent class model can be expressed as follows:

$$P(\mathbf{Y} = \mathbf{y} | X = x) = \sum_{c=1}^{C} \gamma_c(x) \prod_{j=1}^{J} \prod_{r_j=1}^{R_j} \rho_{j,r_j|c}^{I(y_j=r_j)}, \tag{6.1}$$

where $\gamma_c(x) = P(L = c | X = x)$ is a standard baseline-category multinomial logistic model (e.g., Agresti, 1990).

With a single covariate X, $\gamma_c(x)$ can be expressed as follows:

$$\gamma_c(x) = P(L = c | X = x) = \frac{e^{\beta_{0c}+\beta_{1c}x}}{1 + \sum_{c'=1}^{C-1} e^{\beta_{0c}+\beta_{1c}x}} \tag{6.2}$$

for $c' = 1, ..., C - 1$. Logistic regression requires designating one category of the criterion variable as the reference category. In this notation the designated reference category is latent class C. The choice of reference category is arbitrary and will not affect the results in any substantive way, but it can affect the ease of interpretation of the results.

To avoid making the notation overly complicated, we have presented these equations with a single covariate; however, multiple covariates can be incorporated. When there is more than one covariate we will refer to the slopes as β_{1c} for the first covariate, β_{2c} for the second covariate, and so on. When there is an interaction between two covariates the slope will be designated with both subscripts; for example, the interaction of covariates X_1 and X_2 is denoted β_{12c}. An example with more than one covariate will be presented in Section 6.8.

The logistic regression analysis produces an estimate of the effect for each latent class in comparison to the reference latent class. Thus, there will be $C - 1$ regression coefficients β_{1c} corresponding to each covariate, plus $C - 1$ intercepts β_{0c}. For example, if there were three covariates, there would be three regression coefficients plus an intercept corresponding to each latent class except the one designated as the reference category.

6.4.1   What is estimated

As discussed in previous chapters, in a traditional LCA two sets of parameters are estimated: the item-response probabilities (ρ's) and the latent class prevalences (γ's). In LCA with covariates, the item-response probabilities are still estimated, but not the latent class prevalences. Instead of the latent class prevalences, regression coefficients (β's) are estimated, and the latent class prevalences can be expressed as functions of the regression coefficients and individuals' values on the corresponding covariates. The latent class prevalences are still of conceptual interest.

6.4.2   Treatment of item-response probabilities in LCA with covariates

It is important to note that in the model expressed in Equations 6.1 and 6.2 the covariate X can be related to the latent class prevalences (i.e., the γ's) but not to the item-response probabilities (i.e., the ρ's). This means that measurement invariance across all values of any covariates is assumed implicitly. If this assumption is violated, the model is misspecified, and the results may be misleading.

It is possible to fit models in which covariates are related to both the latent class prevalences and the item-response probabilities, but the models are challenging to interpret because of their complexity. They also require estimation of many parameters and therefore can be unstable. However, there may be times when the research question of interest requires such models. Humphreys and Janson (2000) provided an interesting example of this. These models are outside the scope of this book.

6.5   HYPOTHESIS TESTING IN LCA WITH COVARIATES

Hypothesis testing in LCA with covariates is done by means of a likelihood ratio χ^2 test. Examples of this will be seen below; in this section we provide a general introduction to hypothesis testing in LCA with covariates.

Suppose that a latent class model involves a single covariate X. The null hypothesis $H_0 : \beta_1 = 0$ (where β_1 refers to the vector of $C - 1$ regression coefficients corresponding to the covariate X) can be tested by comparing the fit of two models. Model 1 is the baseline model without the covariate X, estimating p_1 parameters. Model 2 is the corresponding model that includes X, estimating p_2 parameters. Then $-2(\ell_1 - \ell_2)$ is in theory distributed as a χ^2 with $df = p_2 - p_1$. Note that in this case

$p_2 - p_1 = C - 1$, the number of regression coefficients associated with the covariate X.

In models with two or more covariates it is often necessary to evaluate the statistical significance of one particular covariate. Here the null hypothesis is that the covariate of interest does not contribute significantly to the prediction of latent class membership over and above the contribution of the other covariates in the model. This hypothesis can be tested by comparing the fit of a model that includes all of the covariates against that of a corresponding model that includes all of the covariates except the particular covariate being tested. To be more specific, suppose that a latent class model includes K covariates, where $k = 1, ..., K$. The statistical significance of a single covariate X_k can be tested by means of a likelihood-ratio test that compares two models. In Model 1, all covariates are included except X_k; this model estimates p_1 parameters. In Model 2, all K covariates are included; this model estimates p_2 covariates. Then the hypothesis test proceeds in much the same way as described in the preceding paragraph; $-2(\ell_1 - \ell_2)$ is in theory distributed as a χ^2 with $df = p_2 - p_1 = C - 1$, where again $C - 1$ = the number of regression coefficients associated with the covariate X.

Sometimes it is desired to test the significance of a set or block of covariates rather than a single covariate. This can be important, for example, when two or more dummy-coded (0/1) variables are used to represent a single categorical covariate with three or more response options. In this case the overall test of significance of that covariate requires testing the significance of the entire set of dummy-coded variables. This can be accomplished by using essentially the same strategy as has been outlined above, except that the difference between Models 1 and 2 involves the entire block of covariates. For example, suppose that an omnibus test of the significance of all K covariates in a latent class model is desired. In this case, Model 1 would be the baseline model with no covariates, and Model 2 would be the model with the entire set of K covariates included. Here $df = p_2 - p_1 = K(C - 1)$.

6.6 INTERPRETATION OF THE INTERCEPTS AND REGRESSION COEFFICIENTS

6.6.1 Understanding odds and odds ratios

To interpret the regression coefficients in logistic regression it is necessary to have an understanding of odds and odds ratios. The odds of Event 1 in relation to Event 2 are the probability of Event 1 divided by the probability of Event 2. The simplest example of odds is when there are only two possibilities, as for example when the event is a coin coming up heads. Then Event 1 is occurrence of the event (heads) and

Event 2 is nonoccurrence of the event (not heads, or tails). If Event 1 has probability $P(Event\ 1)$ of occurring, the expression for the odds is

$$odds = \frac{P(Event\ 1)}{1 - P(Event\ 1)}. \tag{6.3}$$

For example, assuming that the coin is fair, the probability of heads $P(Event\ 1) = .5$. The odds of a fair coin coming up heads in a single toss is $.5/.5 = 1.0$. This means that heads and tails are equally likely to occur. Odds are often expressed as "numerator-to-one"; in this example, the odds of heads occurring in a single toss of a fair coin might be said to be one-to-one.

More generally, odds can be used to express the relative chances of any two events:

$$odds = \frac{P(Event\ 1)}{P(Event\ 2)}. \tag{6.4}$$

The first line in Table 6.1 shows the overall prevalences of each of the latent classes in the health behavior example. The odds of being in the Healthy latent class relative to being in the Typical latent class can be expressed as $.27/.37 = .73$. In other words, a randomly selected individual is 73 percent as likely to be in the Healthy latent class as in the Typical latent class.

Scientific questions concerning whether odds are related to a covariate, such as group membership, can be examined in terms of the *odds ratio*. For example, differences in probabilities across genders can be expressed in an odds ratio as follows:

$$odds\ ratio = \frac{\frac{P(Event1|male)}{P(Event2|male)}}{\frac{P(Event1|female)}{P(Event2|female)}}. \tag{6.5}$$

The odds ratio ranges from 0 to infinity. An odds ratio of 1 would mean that the odds are equal for males and females, in other words, the relative event probabilities are independent of gender. An odds ratio greater or less than 1 would imply that the odds are different in males and females, in other words, the relative event probabilities are related to gender. It is arbitrary whether the odds for males or females appears in the denominator of Equation 6.5, and this is typically determined by the coding of the covariate. It would have been equally valid to express the odds ratio with males in the denominator. With females in the denominator, an odds ratio > 1 means that the odds of Event 1 relative to Event 2 are greater for males; an odds ratio < 1 means that the odds of Event 1 relative to Event 2 are greater for females.

To illustrate the use of the odds ratio, we will return to the multiple-group model that was fit to the health behavior data in Chapter 5 to examine gender differences. The latent class prevalences for this model appear in the lower part of Table 6.1. Suppose that we are interested in seeing whether the odds of being in the Healthy latent class relative to being in the Typical latent class are different for males and females. If gender is coded 1 for male and 0 for female, the corresponding odds ratio would be

$$\frac{\frac{P(Healthy|male)}{P(Typical|male)}}{\frac{P(Healthy|female)}{P(Typical|female)}} = \frac{\frac{.27}{.42}}{\frac{.15}{.28}} = 1.20. \tag{6.6}$$

This odds ratio of 1.20 can be interpreted as follows: The odds of being in the Healthy latent class as compared to the Typical latent class are 1.20 times greater for males than for females; or, the odds are about 20 percent greater for males than for females.

6.6.2 The correspondence between regression coefficients and odds/odds ratios

The intercepts (β_0's) in a logistic regression can be transformed into odds, and the other regression coefficients (β_1's, β_2's, etc.) can be transformed into odds ratios by exponentiating the coefficients (i.e., e^{β_0} = odds and e^{β_1} = odds ratios). For many people this is the most intuitively appealing way to interpret these quantities.

In fact, an alternative form of Equation 6.2 can be found by first expressing γ_c in terms of odds. Let latent class C be the reference category. Then the odds of latent class c in relation to the reference latent class C are expressed as

$$\frac{\gamma_c}{\gamma_C}. \tag{6.7}$$

The log of the odds is often called the *logit*:

$$logit = \log\left(\frac{\gamma_c}{\gamma_C}\right). \tag{6.8}$$

Then Equation 6.2 can be expressed in terms of the logit, as follows:

$$\log\left(\frac{\gamma_c}{\gamma_C}\right) = \beta_{0c} + \beta_{1c}x. \tag{6.9}$$

The health behavior example using gender as a covariate can be used as an example to show why the intercepts in a logistic regression can be expressed as odds and the regression coefficients in a logistic regression can be expressed as odds ratios.

Let us begin with the intercept. As mentioned above, in the example gender is coded so that for males, $X = 1$ and for females $X = 0$. Substituting these values for X in Equation 6.9 shows that the log odds of membership in the Healthy latent class for females, using Typical as a reference class, is

$$\log\left(\frac{\gamma_{Healthy}}{\gamma_{Typical}}\right) = \beta_{0Healthy} + \beta_{1Healthy}(0) = \beta_{0Healthy}. \tag{6.10}$$

Thus exponentiating the intercept corresponding to the Healthy latent class, $\beta_{0Healthy}$, produces the odds of membership in the Healthy latent class in relation to the Typical latent class for females:

$$e^{\beta_{0Healthy}} = \frac{\gamma_{Healthy}}{\gamma_{Typical}}. \tag{6.11}$$

More generally, $e^{\beta_{0c}}$ represents the odds of membership in latent class c in relation to the reference latent class C when $X = 0$.

Now let us show how exponentiating a regression coefficient produces an odds ratio. The log odds of membership in the Healthy latent class for males is

$$\log\left(\frac{\gamma_{Healthy}}{\gamma_{Typical}}\right) = \beta_{0Healthy} + \beta_{1Healthy}(1). \tag{6.12}$$

The log of the odds ratio is

$$\log\left(\frac{\frac{P(Healthy|male)}{P(Typical|male)}}{\frac{P(Healthy|female)}{P(Typical|female)}}\right) = \log\left(\frac{P(Healthy|male)}{P(Typical|male)}\right) - \log\left(\frac{P(Healthy|female)}{P(Typical|female)}\right) \tag{6.13}$$

$$= \beta_{0Healthy} + \beta_{1Healthy}(1) - \beta_{0Healthy} - \beta_{1Healthy}(0) \tag{6.14}$$

$$= \beta_{1Healthy} \tag{6.15}$$

and therefore

$$e^{\beta_{1Healthy}} = \frac{\frac{P(Healthy|male)}{P(Typical|male)}}{\frac{P(Healthy|female)}{P(Typical|female)}}. \tag{6.16}$$

In this example, $e^{\beta_{1Healthy}}$ represents the ratio of the odds of membership in the Healthy latent class in relation to the Typical latent class for males compared to females. **More generally, the effect of a covariate X, $e^{\beta_{1c}}$, reflects the change in odds of membership in latent class c in relation to the reference latent class C associated with a one-unit change in X.**

It is important to note that in the paragraph above we are using the word *effect* in a statistical sense, not a causal sense. When we say that the effect of a one-unit change in X is, say, a doubling of a particular odds ratio, we do not mean that if it were possible to manipulate a one-unit change in X experimentally, a doubling of the odds ratio would follow as a consequence. Instead, we are using the term *effect* to convey the idea that the model we have fit suggests than when a one-unit change in X is observed, a doubling of the odds ratio is observed as well. In this book we have done our best to avoid language implying that the predictor variables have a causal relationship with latent class membership. There may be instances in which causation may be inferred, for example, if the covariate is a dummy variable expressing random assignment to experimental conditions. However, in general it is not appropriate to infer causation from an ordinary regression equation. Some new work is opening up possibilities for causal inference; we refer the reader to Gelman and Meng (2004), Morgan and Winship (2007), Robins, Hernán, and Brumback (2000), Rubin (2005), and Schafer and Kang (2008).

6.7 EMPIRICAL EXAMPLES OF LCA WITH A SINGLE COVARIATE: POSITIVE HEALTH BEHAVIORS

In this section, we present two separate analyses each incorporating a single co-variate into the positive health behavior example. In the first analysis we include a dummy-coded variable, gender, as the covariate. In the second analysis we include a numeric variable, maternal education, as the covariate. In Section 6.8 we include both covariates in a single analysis.

6.7.1 Results of logistic regression using gender as a covariate

Table 6.3 shows the results of an analysis using gender as a covariate to predict membership in positive health behavior latent classes. As described in Section 6.3.1, we dummy-coded the gender variable so that a 1 represented males and a 0 represented females.

The ℓ for this model appears at the bottom of the table. As discussed in Section 6.5, the hypothesis test for the effect of gender is conducted by comparing the ℓ for the baseline model with no covariates (see Table 6.1) to the ℓ for the model that included gender as a covariate. Thus the hypothesis test was $-2(\ell_1 - \ell_2) = -2(-9,533.1 - (-9,501.8)) = 62.6$. There were four degrees of freedom, because the model that included gender as a covariate estimated four more parameters than

Table 6.3   Gender as a Predictor of Membership in Latent Classes of Positive Health Behaviors (Monitoring the Future Data, 2004; $N = 2,065$)

	Latent Class				
	Unhealthy	Sleep Deprived	Typical	Healthy Eaters	Healthy
Intercepts					
β_0's	−1.21	−.16	ref	−.15	−.65
Odds	.30	.86	ref	.86	.52
Gender ($p < .0001$)					
β_1's	−.04	−1.21	ref	−1.52	.20
Odds ratios	.96	.30	ref	.22	1.23

$\ell = -9,501.8$.

the baseline model. The hypothesis test was significant at $p < .0001$, indicating that gender was a statistically significant predictor of latent class membership. In other words, the distribution of latent class membership differed across gender.

Table 6.3 shows the estimates of the intercepts (β_0's) and their corresponding transformation to odds, and the regression coefficients (β_1's) and their corresponding transformations to odds ratios. The Typical latent class served as the reference latent class. All of the intercepts had a negative sign. This reflects the fact that when $X = 0$, that is, for females, each of the corresponding latent classes had a smaller prevalence than the reference latent class. This is also reflected in the odds corresponding to the intercepts, which were less than 1 for each of the latent classes. For example, the odds for the Healthy latent class were .52, implying that females were about half as likely to be in the Healthy latent class as they were to be in the Typical latent class.

The sign of the coefficients for the dummy-coded gender variable reflects which gender has a higher probability of membership in the corresponding latent class relative to the Typical latent class. These signs indicated that males had a higher probability than females of membership in the Healthy latent class, and a lower probability of membership in the Unhealthy, Sleep Deprived and Healthy Eaters latent classes, relative to membership in the Typical latent class. This is consistent with the latent class prevalences for each gender shown in Table 6.1. As discussed above, the odds ratios offer a direct comparison of these relative probabilities. For example, Table 6.3 shows that for males the odds of being in the Healthy latent class relative to the Typical latent class was approximately 1.23 times the corresponding odds for females. This is identical (within rounding) to the odds ratio computed in Equation 6.6.

Examining the remaining odds ratios and comparing them to the gender differences in latent class prevalences that are displayed in Table 6.1 helps to illustrate the interpretation of the odds ratios. The odds ratio of .96 corresponding to Unhealthy indicates that the odds of being in the Unhealthy latent class relative to the Typical latent class were about even for males and females. This may seem counterintuitive, because it is evident from Table 6.1 that for males the probability of membership in the Unhealthy latent class was considerably larger than it was for females. However, membership in the Typical latent class was also considerably larger for males. Thus the odds of membership in the Unhealthy latent class relative to the Typical latent class were about the same for each gender, producing an odds ratio close to 1. By contrast, the odds ratios corresponding to Sleep Deprived and Healthy Eaters indicate that for males, the odds of being in these latent classes relative to the Typical latent class were considerably lower than these odds are for females. There is no odds ratio for Typical because it is the reference latent class.

6.7.1.1 Effect of changing the designated reference latent class

What would these results have looked like if a different latent class had been selected as the reference class? The hypothesis tests would have been exactly the same. However, the values of individual β_0's, β_1's, odds, and odds ratios would have been different. For example, if the Unhealthy latent class had been used as a reference category, the intercepts would all have been positive, because for females, the probability of membership in any of the other latent classes was larger than the probability of membership in the Unhealthy latent class. There would have been no coefficient for Unhealthy, but there would have been one for Typical. Both the odds and the odds ratio for Typical using Unhealthy as the reference latent class would have been simply the inverse of the odds ratio for Unhealthy using Typical as the reference latent class. Thus, as discussed above, the choice of the reference latent class has no impact on the results beyond affecting the direction and, in some cases, ease of interpretation.

6.7.2 Results of logistic regression using maternal education as a covariate

Table 6.4 shows the results of an analysis using maternal education as a covariate to predict membership in positive health behavior latent classes. The maternal education variable had six categories of education, ranging from "Grade school" (originally coded as 1) to "Graduate school" (originally coded as 6), plus a code for "Don't know." We recoded the "Don't know" category as missing and then standardized the maternal education variable before using it as a covariate. As in the analysis with gender as a covariate, the Typical latent class was designated as the reference latent class.

The hypothesis test for the significance of the maternal education covariate was conducted in much the same way as the hypothesis test for the gender covariate described in Section 6.7.1. The ℓ for the baseline model with no covariates (see Table 6.1) was compared to the ℓ for the model that included maternal education as a covariate (the ℓ appears at the bottom of Table 6.4). Thus the hypothesis test was $-2(\ell_1 - \ell_2) = -2(-9,533.1 - (-9,501.4)) = 63.4$, with $df = 4$. This hypothesis test was significant at $p < .0001$, indicating that maternal education was a statistically significant predictor of latent class membership.

Table 6.4 shows the estimates of the intercepts (β_0's), odds, regression coefficients (β_1's), and odds ratios. The Typical latent class served as the reference latent class. All of the intercepts were negative. This means that individuals for whom maternal

Table 6.4   Maternal Education as a Predictor of Membership in Latent Classes of Positive Health Behaviors (Monitoring the Future Data, 2004; $N = 2,065$)

	Latent Class				
	Unhealthy	Sleep Deprived	Typical	Healthy Eaters	Healthy
Intercepts					
β_0's	−.90	−1.15	ref	−1.10	−.35
Odds	.41	.32	ref	.33	.70
Maternal education ($p < .0001$)					
β_1's	−.22	−.03	ref	.12	.38
Odds ratios	.80	.98	ref	1.12	1.47

$\ell = -9,501.4.$

education equaled zero (which is the mean of this standardized variable) were less likely to be in the Unhealthy, Sleep Deprived, Healthy Eaters, or Healthy latent classes than they were to be in the Typical latent class.

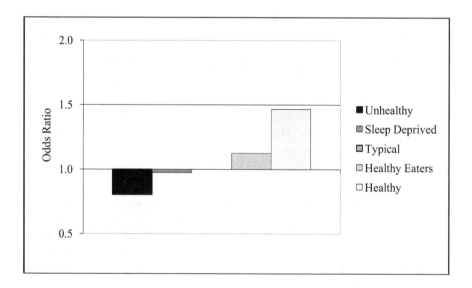

Figure 6.1   Overall effect of one-standard-deviation increase in maternal education in positive health behavior example (Monitoring the Future data, 2004; $N = 2,065$). The Typical latent class is the reference group.

Figure 6.1 is a graphical representation of the effect of the maternal education covariate on membership in positive health behavior latent classes. As Table 6.4 shows, the regression coefficients corresponding to the Healthy Eaters and Healthy latent classes were positive. This means that a one-unit [i.e., a one-standard-deviation (because maternal education is standardized)] increase in maternal education was associated with an increase in the odds of being in each of these latent classes relative to the Typical latent class. For example, if maternal education was increased by one standard deviation, the odds of being in the Healthy latent class relative to the Typical latent class changed by a factor of 1.47; in other words, the odds increased by nearly 50 percent. The regression coefficients corresponding to Unhealthy and Sleep Deprived were negative. This means that a one-unit (i.e., a one-standard-deviation) increase in maternal education was associated with a *decrease* in the odds of being in each of these latent classes relative to the Typical latent class. For example, if maternal education was increased by one standard deviation, the odds of being in the Unhealthy latent class relative to the Typical latent class changed by a factor of .80; in other words, the odds decreased. Overall, the results suggest that better educated mothers were more likely to have adolescents in the latent classes that reflected better health and less likely to have adolescents in the latent classes that reflected poorer health, as compared to the Typical latent class.

6.8   EMPIRICAL EXAMPLE OF MULTIPLE COVARIATES AND INTERACTION TERMS: POSITIVE HEALTH BEHAVIORS

Models with more than one covariate enable examination of the effects of each covariate while the effects of other covariates have been partialed out. They also open up the possibility of examining interactions between covariates. To demonstrate this, let us examine both gender and maternal education as covariates in the positive health behavior example. We also add the interaction between gender and maternal education.

Interactions may be added into LCA with covariates using the approach generally used in regression described in Aiken and West (1991) and Jaccard, Turrisi, and Wan. (2003). Some software programs allow the user to specify interactions by listing variables that interact. Others require the user to compute interactions by multiplying variables together and including the resulting product as a covariate. When interactions between variables are to be examined in regression, it is recommended (e.g., Aiken and West, 1991) that they be centered (i.e., the mean should be subtracted) first unless they are dummy coded.

To compute the interaction between gender and maternal educations in our example, we first standardized maternal education, which took care of the requirement that the variable be centered. Then we computed a new product variable, gender

Table 6.5   Gender and Maternal Education as Predictors of Membership in Latent Classes of Positive Health Behaviors (Monitoring the Future Data, 2004; $N = 2,065$)

	Latent Class				
	Unhealthy	Sleep Deprived	Typical	Healthy Eaters	Healthy
Intercept					
β_0's	−1.22	−.19	ref	−.22	−.62
Odds	.29	.83	ref	.80	.54
Gender					
β_1's	−.04	−1.22	ref	−1.62	.16
Odds ratios	.96	.29	ref	.20	1.18
Maternal Education					
β_2	−.01	−.03	ref	.38	.64
Odds ratios	.99	.97	ref	1.46	1.89
Gender X Maternal Education					
β_{12}'s	−.36	−.37	ref	−.71	−.47
Odds ratios	.69	.69	ref	.49	.63

$\ell = -9,464.0$.

× maternal education (standardized). The regression thus included three covariates: gender, maternal education (standardized), and the product variable. Note that co-variates should be standardized before using them to compute a product, and that the resultant product should not then be standardized before introducing it as a covariate.

The results of the analysis using gender, maternal education, and gender × maternal education as covariates are shown in Table 6.5. The β's corresponding to the intercept and to the main effects of gender and maternal education are a bit different from their counterparts in Tables 6.3 and 6.4. This is due to the presence of additional covariates in the model in Table 6.5.

Table 6.5 shows that the ℓ for the overall model is −9,464.0. Each hypothesis tests for a covariate was performed by comparing this ℓ to that of a corresponding model, omitting the covariate. The details of these hypothesis tests are shown in Table 6.6. The hypothesis tests show that gender and maternal education, and their interaction, are all statistically significant.

Table 6.6   Hypothesis Tests for Gender, Maternal Education, and Their Interaction for Model of Positive Health Behaviors Reported in Table 6.5 (Monitoring the Future Data, 2004; $N = 2,065$)

Covariate	ℓ Removing Covariate	Likelihood-Ratio Statistic	df	p
Gender	−9,492.3	56.6	4	<.0001
Maternal Education	−9,488.8	49.7	4	<.0001
Interaction	−9,470.6	13.2	4	.01

6.8.1 Interpretation of the interaction between gender and maternal education

Table 6.5 shows that the odds ratios corresponding to each of the latent classes were quite a bit less than 1 for the gender × maternal education interaction, and Table 6.6 shows that the interaction covariate was significant. How can this be interpreted?

Statistical interactions are always challenging to interpret. This is particularly true in regression, and most particularly true in logistic regression, because odds ratios corresponding to interactions are complex. For this reason we will spend some additional time on considering the interaction between gender and maternal education. In general, we recommend taking a graphical approach that enables examination of the effects of one variable while the other is held constant. In our example this can be accomplished by plotting the effects of maternal education separately by gender.

An interaction between two variables means that the effect of one variable is different depending on the level of the other variable. In the example, the significant interaction between gender and maternal education means that the effect of maternal education on health behavior latent class membership differed in some way between males and females. Figure 6.2 plots the odds ratios corresponding to a one-standard-deviation increase in maternal education for males and for females. For both genders

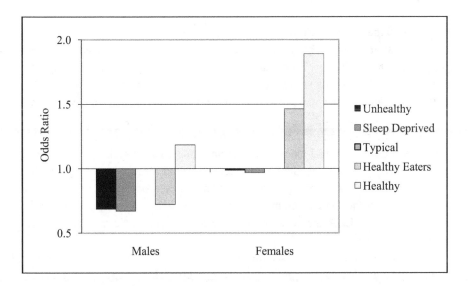

Figure 6.2 Effect of one-standard-deviation increase in maternal education for males and females in positive health behavior example (Monitoring the Future data, 2004; $N = 2,065$).

an increase in maternal education was associated with an increase in the likelihood of membership in the Healthy latent class in relation to the Typical latent class. This increase was considerably higher for females than for males. The relationship between maternal education and the odds of membership in the Healthy Eaters latent class in relation to the Typical latent class was opposite for the two genders. For girls, an increase in maternal education was associated with an increase in the odds of membership in the Healthy Eaters latent class; for boys, higher maternal education was associated with a decrease in the odds of membership in this latent class compared to the Typical latent class. For girls, an increase in maternal education was not associated with a change in the odds of membership in the Sleep Deprived or Unhealthy latent classes compared to the Typical latent class; for boys, an increase in maternal education was associated with a decrease in these odds. Thus, the overall pattern of the relation of maternal education to latent class membership differed for males and females. Higher maternal education had a positive association with better health behaviors, particularly among females, whereas lower maternal education appeared to be a risk factor for unhealthy behavior among boys only.

6.9 MULTIPLE-GROUP LCA WITH COVARIATES: MODEL AND NOTATION

In Chapter 5 the positive health behavior data set was modeled using multiple-group LCA. In that analysis the groups (i.e., males and females) were treated as separate populations, and parameter estimates were conditioned on group membership. It is possible to use a grouping variable approach while incorporating a covariate. In this approach the intercepts and regression coefficients for all covariates are conditioned on the grouping variable.

Thus an alternative to including gender, maternal education, and their product as covariates in our empirical example is to retain maternal education as a covariate, but instead of including gender as a covariate, include it as a grouping variable. Then regression weights for maternal education will be produced separately for the male group and the female group. In other words, when a covariate is included in multiple-group LCA the covariate and the grouping variable are automatically allowed to interact.

Equations 6.1 and 6.2 expressed the latent class model with a covariate. Now let us add a grouping variable V with $q = 1, ..., Q$ groups. The corresponding equations when the model includes both a grouping variable V and a covariate X are

$$P(\mathbf{Y} = \mathbf{y}|V = q, X = x) = \sum_{c=1}^{C} \gamma_{c|q}(x) \prod_{j=1}^{J} \prod_{r_j=1}^{R_j} \rho_{j,r_j|c,q}^{I(y_j=r_j)}. \tag{6.17}$$

Then $\gamma_{c|q}(x) = P(L = c|V = q)$ is a standard baseline-category multinomial logistic model (e.g., Agresti, 2007). With a single covariate X, $\gamma_{c|q}(x)$ can be expressed as follows:

$$\gamma_{c|q}(x) = P(L = c|V = q) = \frac{e^{\beta_{0c|q} + \beta_{1c|q}x}}{1 + \sum_{c'=1}^{C-1} e^{\beta_{0c|q} + \beta_{1c|q}x}} \tag{6.18}$$

for $c' = 1, ..., C - 1$, where C is the reference latent class. The use of the subscript q in the intercepts $\beta_{0c|q}$ and slopes $\beta_{1c|q}$ indicates that these parameters are conditioned on group membership. Note that in this model, the item-response probabilities may be conditioned on the grouping variable, but they may not be conditioned on the covariate.

6.9.1 Empirical example: Positive health behaviors

Table 6.7 contains the results of the LCA, including maternal education as a covariate and gender as a grouping variable. As Table 6.7 shows, this analysis produced estimates of the intercepts and regression coefficients for maternal education separately for males and females. An overall test of the significance of maternal education was produced by comparing ℓ for the model in Table 6.7 with the ℓ for a model that included gender as a grouping variable but did not include maternal education as a covariate. Because the latter model did not estimate the regression coefficients for maternal education for the two groups, it included eight fewer parameters. The test statistic is thus $-2[-9,501.8 - (-9,464.0)] = 75.6, df = 8, p < .0001$. We defer interpretation of this hypothesis test to Section 6.10.2.

6.10 GROUPING VARIABLE OR COVARIATE?

Certain categorical variables, such as gender, ethnicity, political affiliation, religious affiliation, and treatment group, can be incorporated into LCA as either a grouping variable in a multiple-group LCA or a dummy-coded variable (or set of dummy-coded variables) in LCA with covariates. How should the investigator choose between these alternative approaches? When deciding whether to take a multiple-group or a covariate approach to LCA, it may help to bear in mind some statistical and conceptual differences between them, and also to understand the circumstances under which the two approaches are mathematically equivalent.

Table 6.7   Maternal Education as a Predictor of Membership in Latent Classes of Positive Health Behaviors, with Gender as a Grouping Variable (Monitoring the Future Data, 2004; $N = 2,065$)

	Latent Class				
	Unhealthy	Sleep Deprived	Typical	Healthy Eaters	Healthy
Male					
Intercept					
β_0's	−1.26	−1.41	ref	−1.83	−.45
Odds	.28	.24	ref	.16	.64
Maternal Education					
β_1's	−.38	−.40	ref	−.33	.17
Odds ratios	.69	.67	ref	.72	1.18
Female					
Intercept					
β_0's	−1.23	−.19	ref	−.22	−.62
Odds	.29	.83	ref	.80	.54
Maternal Education					
β_1's	−.01	−.03	ref	.38	.64
Odds ratios	.99	.97	ref	1.46	1.89

$\ell = -9,464.0$.

6.10.1   How the multiple-group and covariate models are different

Multiple-group LCA and LCA with covariates differ conceptually in several ways. First, the multiple-group and covariate approaches reflect different points of view about the grouping variable. In Chapter 5 we discussed how the multiple-group approach starts with the idea that the groups represent distinct populations. From this perspective it is important to conduct a thorough examination of any observed group differences, including differences in the relation between all covariates and membership in the latent classes. Thus in multiple-group LCA, group membership is treated differently from any covariates. In other words, the grouping variable is automatically allowed to moderate the effect of *each* covariate on latent class membership. By contrast, the covariate approach starts with the idea that there is a single population of research participants, but the participants vary in terms of latent class membership, and this variance can be explained by one or more covariates. From this perspective, the group membership variable is just one of the covariates, and is no different from any other.

Second, a model including a grouping variable along with one or more covariates addresses subtly different research questions than an equivalent model that treats the grouping variable as a dummy-coded covariate. The former model addresses the question: How do the intercepts and regression weights vary across groups? The latter addresses the question: What are the effects of each covariate with the effects of the other covariates partialed out? When there are numerous covariates, the latter model can often be more parsimonious, particularly when not every possible interaction with the dummy-coded grouping variable is included in the regression. It is also often much more straightforward to interpret the smaller number of regression weights that are produced by a covariate model. When group differences in the effect of a covariate are of interest, an interaction term can be coded between the covariate and the dummy-coded grouping variable, as demonstrated above. However, if an interaction between a covariate and the dummy-coded grouping variable is not coded, group differences in the effect of the covariate may be undetected. Thus the covariate approach carries with it a risk of failing to uncover important group differences that are not explicitly modeled.

Third, the multiple-group approach does not require the assumption of measurement invariance. In the multiple-group approach the investigator has the option of allowing some or all of the item-response probabilities to vary across groups (see Chapter 5). This option is not available in the covariate approach presented in this chapter (again, for another perspective, see Humphreys and Janson, 2000); in other words, the covariate approach requires the assumption of measurement invariance.

6.10.2 When the multiple-group and covariate models are mathematically equivalent

Suppose that (1) two latent class models involve the same observed variables and the same number of latent classes; (2) there is a categorical variable that could be either a grouping variable or a covariate; and (3) the models are to be fit to the same data. In Model 1 the categorical variable is treated as a grouping variable, and the item-response probabilities are constrained to be equal across groups. In Model 2 the categorical variable is dummy coded and treated as a covariate (or a set of covariates).

These two models estimate the same number of parameters. The number of item-response probabilities is identical; Model 1 estimates a single set of item-response probabilities that holds across all groups, and Model 2 does essentially the same, because it does not identify any groups. The number of latent class prevalences in Model 1 is equal to the number of regression coefficients estimated in Model 2. In Model 1, in which latent class prevalences are conditioned on the group, there are $C - 1$ latent class prevalences estimated for each group, for a total of $Q(C - 1)$ latent class prevalences estimated. In Model 2, a categorical variable with Q categories is entered into the model as $Q - 1$ dummy variables. Because one latent class is designated as the reference latent class, this results in estimation of $(Q - 1)(C - 1)$ regression coefficients. With the addition of the $C - 1$ intercepts, this makes a total of $Q(C - 1)$ β parameters. Moreover, the latent class prevalence estimates from Model 1 are a simple reparameterization of the intercepts and regression coefficients estimated in Model 2; the latent class prevalence estimates in Model 1 can be used to compute the intercepts and regression coefficients in Model 2, and vice versa. Thus, Models 1 and 2 are mathematically equivalent.

An example of this equivalence can be found by comparing the multiple-group LCA model fit to the health behavior example in Chapter 5 and reported briefly in Table 6.1 (reported fully in Table 5.15) with the covariate model fit to the same data and reported in Table 6.3. The two models have identical likelihood-ratio statistics of –9,501.8, and the odds ratios in Table 6.3 can be reproduced exactly (within rounding error) by using the latent class prevalence estimates for each gender that appear in Table 6.1. This was demonstrated in Equation 6.6, in which the odds ratio corresponding to $\beta_{1Healthy}$ was obtained from the latent class prevalence estimates in Table 6.1.

In much the same way, a latent class model that includes a covariate X, a grouping variable V, and constrains the item-response probabilities to be equal across groups, is identical to a corresponding latent class model that includes the covariate X, the grouping variable V as a dummy-coded covariate, and the $X \times V$ interaction. The two models have identical log-likelihoods and df. When V is treated as a grouping variable the model estimates $C - 1$ regression weights and $C - 1$ intercepts for each group, for a total of $2Q(C - 1)$ parameters estimated. When V is treated as a covariate, there are $C - 1$ intercepts estimated, $(Q - 1)(C - 1)$ regression weights for the dummy-coded grouping variable, $C - 1$ for the covariate, and $(Q - 1)(C - 1)$ for the interaction, for a total of $2Q(C - 1)$ β parameters estimated.

Readers can see an example of this correspondence by comparing the results in Tables 6.5 and 6.7. The intercepts reported in Table 6.5 are identical to the intercepts for the female group shown in Table 6.7. Similarly, the regression coefficients corresponding to the main effect of maternal education reported in Table 6.5 are identical to the regression coefficients for the maternal education covariate for females in Table 6.7. This is because the dummy variable for gender was coded 0 for females and 1 for males; in other words, females were coded as the reference group. Because of the equivalence of the two models, the data for Figure 6.2, which depicts odds ratios separately for males and females in order to illustrate the interpretation of the interaction between gender and maternal education reported in Table 6.5, could have been drawn from either Table 6.5 (i.e., the odds ratios could have been computed for males and females based on the regression coefficient for the interaction between gender and maternal education) or Table 6.7 (which contains the odds ratios associated with maternal education for each gender). These tables show that the two models had identical log-likelihoods and df.

Although the two overall models are identical, they are parameterized differently, so the specific hypothesis tests in a latent class model that treats V as a dummy-coded covariate and includes it with other covariates are subtly different from those in a corresponding model that treats V as a grouping variable. This brings us back to the interpretation of the hypothesis test associated with the model reported in Table 6.7, mentioned in Section 6.9.1. In the model in Table 6.7, the hypothesis test for the overall effect of maternal education was a test of the *combination* of maternal education and the gender × maternal education interaction. The hypothesis test indicated that this combined effect was significant, but it did not indicate which part was significant. By contrast, in the model in Table 6.5, in which gender is included as a dummy-coded covariate, there were separate hypothesis tests for maternal education and the gender × maternal education interaction.

6.11 USE OF A BAYESIAN PRIOR TO STABILIZE ESTIMATION

Sparseness was first mentioned in Section 4.3.2.3, where we explained that the term *sparseness* refers to very small cell counts in a contingency table. We explained that sparseness can cause problems with model selection, model identification, and parameter estimation in LCA.

Sparseness can cause estimation problems in logistic regression as well, irrespective of whether the outcome variable is latent. A maximum likelihood solution for logistic regression parameters is available only if the number of successes and failures is nonzero for all possible combinations of responses on the covariates. For example, for a model in which gender is used to predict success or failure, both the number of successes and the number of failures must be nonzero for males and for females (i.e., at least one male must have reported a success, at least one male must have reported a failure, at least one female must have reported a success, and at least one female must have reported a failure). When this condition is not met, estimation is unstable because odds ratios can be infinite.

In traditional logistic regression with observed variables, sparseness is more likely to cause estimation problems when the overall sample size is small, when one or more categories on the outcome variable has a small count, or when membership in one or more categories on the outcome variable is essentially zero for some level of a covariate. Similarly, in LCA, sparseness is more likely to cause estimation problems when the sample size is small, when one or more groups is small, when one of the latent classes has a very small prevalence, or when membership in one of the classes is essentially zero for some level of a covariate. This last condition can be difficult for users to identify.

One approach that can be helpful when sparseness issues arise in logistic regression is the use of a data-derived Bayesian prior, as described by Clogg, Rubin, Schenker, Schultz, and Weidman (1991). In the Clogg et al. approach, a very small amount of prior information (sometimes referred to as *pseudo-data*) is added to the observed data in a way that either has no impact (when sparseness is not an issue) or biases the mean estimate for each pattern of covariate responses very slightly toward the overall success probability in order to reduce sampling variability. This guarantees that the number of successes and the number of failures is nonzero for each observed combination of responses to the covariates. The prior is referred to as a *flattening* prior because it dampens very slightly the effects of covariates in the observed data in order to make parameter estimation possible. In LCA with covariates, this prior information has no effect on the measurement model, but serves only to stabilize the logistic regression parameters. The flattening prior represents one practical solution for stabilizing the estimation of logistic regression parameters when one or more of the beta estimates diverge to infinity due to insufficient information for estimation. [This stabilizing prior is available as an option in Proc LCA (Lanza, Collins, Lemmon, and Schafer, 2007).]

In multiple-group models with covariates, it is important to establish that no latent class prevalence is estimated at zero for any of the groups. To check this, the latent class prevalences can be examined in the model without covariates, fit only to individuals who provided data on the covariate(s). For each group, any latent class prevalences that are estimated at a value very close to 0 should be fixed to 0 prior to incorporating covariates. This eliminates the empty latent class from the logistic regression for that group.

In practice, if estimation of a logistic regression model fails in LCA with covariates, the problem is most likely due to sparseness in the data. The application of the Bayesian prior described above will solve many sparseness-related estimation problems. In extreme cases, this approach may not suffice and additional measures must be taken. One option is to switch from a baseline-category multinomial logistic regression model to a binomial logistic regression model, described in Section 6.12.

6.12 COMBINING LATENT CLASSES TO APPLY BINOMIAL LOGISTIC REGRESSION IN LCA

Up to this point we have been discussing multinomial logistic regression for predicting latent class membership. In this section we discuss combining latent classes into two aggregate categories and using logistic regression to predict membership in the aggregate categories. When the outcome has only two categories, multinomial logistic regression is often referred to as *binomial logistic regression* .

Why would an investigator consider combining latent classes? One reason might be to highlight a particular research question. For example, suppose that in the positive health behavior example an investigator is interested in what predicts membership in the Healthy latent class compared to membership in any other latent class. In this case a binomial logistic regression that focuses exclusively on the distinction between the Healthy latent class and all of the others combined might be preferred. Another reason might be to deal with estimation problems. As described in Section 6.11, estimation problems can occur when covariates are introduced into LCA in which one or more of the latent classes are small. These problems can often be eliminated by combining latent classes and taking a binomial logistic regression approach.

When an investigator combines latent classes in order to apply binomial logistic regression, one latent class is selected as the *target* latent class. The remaining latent classes are then combined into a single reference latent class, thereby creating an outcome variable with two categories. As discussed above, in multinomial logistic regression with K covariates, $C - 1$ intercepts and $K(C - 1)$ regression weights are estimated, for a total of $(K + 1)(C - 1)$ regression coefficients. By contrast, in binomial logistic regression only one intercept and K regression weights are estimated, for a total of $K + 1$ regression coefficients.

6.12.1 Empirical example: Positive health behaviors

In this example of binomial logistic regression using the positive health behavior data, the Healthy latent class was specified as the target latent class. The remaining four latent classes were then combined into a single reference group. Thus the research questions concerned the likelihood of membership in the Healthy latent class compared to membership in any other latent class.

Table 6.8 shows the results of a binomial LCA in the positive health behavior example, including gender as a covariate (with male coded as 1 and female coded as 0, as in the previous analyses). The positive regression weight indicated that the odds of membership in the Healthy latent class relative to the remaining latent classes were greater for males than for females. As the odds ratio shows, these odds were about twice as large for males as for females. The hypothesis test was conducted by comparing the fit of the model in Table 6.8 against a corresponding model with no covariate. The resulting likelihood-ratio statistic was 7.2, $df = 1, p = .007$, indicating that gender was a significant covariate.

Table 6.9 shows the results of a binomial model that adds maternal education as a covariate along with the gender $\times$ maternal education interaction. Table 6.10

Table 6.8   Binomial Logistic Regression with Gender as a Covariate in Model of Positive Health Behaviors (Monitoring the Future Data, 2004; $N = 2,065$)

Intercept	
β_0	−2.02
Odds	.13
Gender	
β_1	.64
Odds ratio	1.89

$\ell = -9,529.5.$

Table 6.9   Binomial Logistic Regression Model with Gender and Maternal Education as Covariates (Monitoring the Future Data, 2004; $N = 2,065$)

Intercept	
β_0	−1.19
Odds	.30
Gender	
β_1	.25
Odds ratio	1.29
Maternal Education	
β_2	.61
Odds ratio	1.84
Gender X Maternal Education	
β_{12}	−.28
Odds ratio	.76

$\ell = -9,503.1$.

Table 6.10   Hypothesis Tests for Gender, Maternal Education, and Their Interaction for Model in Table 6.9 (Monitoring the Future Data, 2004; $N = 2,065$)

Covariate	ℓ Removing Covariate	Likelihood-Ratio Statistic	df	p
Gender	−9,504.9	3.5	1	.06
Maternal Education	−9,522.7	39.2	1	<.0001
Interaction	−9,505.3	4.3	1	.04

contains the hypothesis tests for the two covariates and their interaction. The effect of gender was marginally significant ($p = .06$), and maternal education ($p < .0001$) and the gender × maternal education interaction ($p = .04$) were both significant.

As we did in a preceding section, we use a figure to help in interpreting the gender × maternal education interaction. Figure 6.3 illustrates the odds ratio associated with a one-standard-deviation increase in maternal education separately for males and females. This figure shows that for males, a one-standard-deviation increase in maternal education was associated with approximately a 40 percent increase in the odds of membership in the Healthy latent class compared to the remaining latent classes; for females, the comparable figure was 84 percent. Thus an increase in maternal education was associated with increased odds of membership in the Healthy latent class for both genders, but the effect was stronger for females.

The model shown in Table 6.11 included gender as a grouping variable and maternal education as a covariate. As would be expected based on the results of the model that included dummy-coded gender and maternal education as covariates (shown in Table 6.9), the model that treated gender as a grouping variable suggested that the effect of maternal education on the odds of membership in the Healthy latent class compared to the remaining latent classes was positive for both genders, and was larger for females than for males. The hypothesis test for the overall effect of maternal education in both genders had likelihood-ratio statistic 51.9, $df = 2$, $p <$

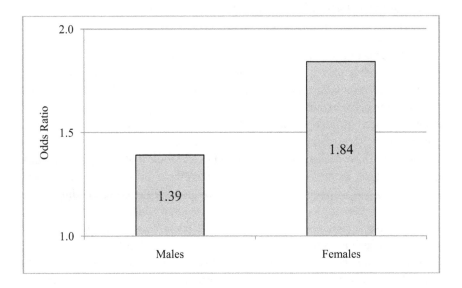

Figure 6.3   Effect of one-standard-deviation increase in maternal education on odds of membership in Healthy latent class as compared to membership in other latent classes combined, by gender (Monitoring the Future data, 2004; $N = 2,065$).

.0001. Note that the odds ratios for males and females reported in Table 6.11 do not match those in Figure 6.3, because they are based on two slightly different models. This is discussed in Section 6.12.2. [Binomial logistic regression for latent class models is available as an option in Proc LCA (Lanza, Collins, Lemmon, and Schafer, 2007).]

Table 6.11   Binomial Logistic Regression with Gender as a Grouping Variable and Maternal Education as a Covariate in Model of Positive Health Behavior (Monitoring the Future Data, 2004; $N = 2,065$)

	Male	Female
Intercept		
β_0	−.88	−1.48
Odds	.41	.23
Maternal Education		
β_1	.30	.69
Odds ratio	1.35	2.00

$\ell = -9,475.8$.

6.12.2   Comparison of binomial multiple groups and covariate models

We noted in Section 6.10.2 that the multinomial logistic regression model with a grouping variable and a covariate is under some circumstances mathematically identical to a corresponding multinomial regression model that includes the covariate, the grouping variable in dummy-coded form, and the interaction of the covariate and the dummy-coded grouping variable. This equivalence does not hold for binomial regression when latent classes have been combined as demonstrated in Section 6.12.1. The reader may have noticed that the log-likelihood for the multiple-group binomial logistic regression model with maternal education as a covariate shown in Table 6.9 (−9,503.1) was different from the log-likelihood for the corresponding model with gender as a dummy-coded grouping variable shown in Table 6.11 (−9,475.8). This is because in binomial regression, the full latent class model with C latent classes is estimated even though for purposes of the regression $C - 1$ latent classes are combined into one. Thus in a multiple-group binomial regression there may be group differences in the prevalence of all C latent classes. By contrast, when the same grouping variable is entered as a dummy-coded covariate into a binomial regression involving combined latent classes, the grouping variable is related to latent class membership only in terms of the target latent class versus all other latent classes.

6.13   SUGGESTED SUPPLEMENTAL READINGS

The scientific literature contains some interesting applications of LCA with covariates. Chung, Flaherty, and Schafer (2006) predicted marijuana use and attitude latent classes using demographic and lifestyle factors, political beliefs, and religiosity. Yamaguchi (2000) predicted membership in three latent classes of gender-role attitudes in a sample of Japanese women. Cleveland, Collins, Lanza, Greenberg, and Feinberg (In press) examined interactions between individual-level risk and contextual-level protective factors as predictors of adolescent substance use. Lanza, Collins, Lemmon, and Schafer (2007) used grades in school and whether or not an adolescent skipped school as covariates to predict an alcohol use latent class. This article also contains information about how to fit LCA models with covariates using Proc LCA.

6.14   POINTS TO REMEMBER

• LCA with covariates is useful for examining predictors of latent class membership. Prediction is done using a logistic regression approach.

• Before conducting LCA with covariates, it is a good idea to establish a baseline model without covariates.

• The model for LCA with covariates shown in this book assumes measurement invariance across all levels of covariates. In this model item-response probabilities cannot be modeled as functions of covariates.

• Logistic regression coefficients can be easier to interpret if they are exponentiated. Then intercepts can be interpreted as odds, and regression coefficients can be interpreted as odds ratios.

• It is necessary to choose one latent class as a reference class for the logistic model used in predicting latent class membership. All odds and odds ratios are computed in relation to the reference class. This choice does not affect hypothesis testing, but it can have an impact on how readily results can be interpreted.

• Interactions between covariates may be included in LCA. All of the usual guidelines about interactions in regression apply. Interpretation of regression coefficients and odds ratios can be complex; figures are often helpful.

• Binomial logistic regression can be used to predict odds of membership in a target latent class compared to a reference category that is an aggregate of all remaining latent classes. This approach may be useful for conceptual or practical reasons.

• In the multinomial logistic regression model there is a direct correspondence between certain multiple-group latent class models and models in which the grouping variable is included as a dummy-coded covariate. This does not necessarily hold for binomial logistic regression when latent classes have been combined.

6.15 WHAT'S NEXT

This chapter concludes the section of the book that focuses on advanced topics in LCA. In the next section we turn to latent class models that are designed especially for analysis of longitudinal data. For the remainder of the book, the emphasis is on modeling change over time. In the next two chapters we discuss repeated-measures LCA (RMLCA), LTA, multiple-group LTA, and LTA with covariates.

PART III

LATENT CLASS MODELS FOR LONGITUDINAL DATA

CHAPTER 7

REPEATED-MEASURES LATENT CLASS ANALYSIS AND LATENT TRANSITION ANALYSIS

7.1 OVERVIEW

In this chapter we introduce approaches to analyzing longitudinal data by means of LCA. Usually when longitudinal data are to be analyzed, the research questions concern some form of change over time. When change is discussed in this book in relation to latent class models, we are referring to change that is in some sense categorical or discrete, such as transitions between latent classes, rather than continuous change in level.

The first approach we discuss is the use of LCA to identify latent classes characterized by different patterns of categorical or discrete change over three or more times. When LCA is applied in this manner it can be called *repeated-measures LCA* (RMLCA). The second approach, *latent transition analysis* (LTA), is a variation of the latent class model that is designed to model not only the prevalence of latent class membership, but the incidence of transitions over time in latent class membership.

7.2 RMLCA

In previous chapters, we discussed approaches to analyzing cross-sectional data using LCA. Under some circumstances, the same kinds of latent class models that have been discussed in these chapters can be fit to longitudinal data using RMLCA. We find that the RMLCA approach works best when a small number of indicators of the latent variable are measured three or more times. Then a latent class model can be fit in which the latent classes correspond to different patterns of categorical or discrete change over time. To illustrate this, we draw on a journal article of ours, Lanza and Collins (2006), which presented a latent class analysis of patterns of heavy drinking in young adults over six time points.

The data are from the 1964 birth cohort of the National Longitudinal Survey of Youth (NLSY). This birth cohort was a nationally representative sample that was established and first measured in 1979 when the participants were 15 years old. The six waves of measurement used in the example presented below correspond to ages 18, 19, 20, 24, 25, and 30. The sample size at the first wave of data collection was $N = 1,265$.

The NLSY included a questionnaire item asking about the frequency of having six or more drinks at one sitting in the last month. The response categories were: "Never," "1 time," "2 or 3 times," "4 or 5 times," "6 or 7 times," "8 or 9 times," and "10 or more times." This variable could be treated as numeric (see Chapter 1) and used to fit a model of continuous growth in heavy drinking, as was done by Muthén and Muthén (2000). However, we were interested in movement in and out of a state of heavy drinking rather than increases and decreases in amount of heavy drinking. Therefore, we decided to recode the variable so that all of the categories except "Never" were combined into a single category reflecting endorsement of having engaged in heavy drinking one or more times in the last month. Because the variable was highly positively skewed at each time point, we felt comfortable combining these categories.

The starting point for the RMLCA was the contingency table formed by cross-tabulating the six dichotomous indicators. The resulting contingency table had $W = 2^6 = 64$ cells. A series of latent class models were fit to this contingency table using the approach described in Chapter 4 and illustrated throughout the book. The model with eight latent classes was judged to fit the data best. The parameter estimates for this model are shown in Table 7.1. The model included some restrictions on the item-response probabilities that were added to achieve better identification.

The six waves of measurement in the NLSY map onto four developmental periods: high school (age 18), post-high school, which will be referred to as college age (age 19–20), young adulthood (age 24–25), and adulthood (age 30). Table 7.2 shows the pattern of heavy drinking across the four developmental periods that characterizes each of the eight latent classes, and Figure 7.1 depicts the patterns graphically. Two

Table 7.1   Eight-Latent-Class Model of Heavy Drinking at Six Different Ages (NLSY; $N = 1,265$) (from Lanza and Collins, 2006)

	Latent Class							
	1	2	3	4	5	6	7	8
Latent class prevalences	.54	.04	.06	.03	.09	.04	.04	.17
Item-response probabilities								
Response = Yes								
Age 18	.07	**.81**[*]	**.81**	.07	.07	.07	.07	**.81**
Age 19	.08	**.74**	**.74**	**.74**	**.74**	.08	.08	**.74**
Age 20	.12	**.82**	**.82**	**.82**	**.82**	.12	.12	**.82**
Age 24	.08	.08	**.85**	.08	**.85**	**.85**	**.85**	**.85**
Age 25	.08	.08	**.76**	.08	**.76**	**.76**	**.76**	**.76**
Age 30	.14	.14	.14	.14	**.77**	.14	**.77**	**.77**

[*] Item-response probabilities > .5 in bold to facilitate interpretation.

of the latent classes are characterized by no change in heavy drinking, two by an increasing trend, two by a decreasing trend, and two by short-term heavy drinking. The most prevalent latent class is characterized by no heavy drinking at any of the developmental periods (54%). The next most prevalent latent class is characterized by heavy drinking across all four developmental periods (17%). These are depicted in the top panel of Figure 7.1. The second panel of Figure 7.1 depicts the two latent classes characterized by an increasing trend in heavy drinking. In one of the latent classes, heavy drinking began in the college-age years (9%), and in the other latent class, heavy drinking began in young adulthood (4%). Individuals in both of these latent classes persisted in heavy drinking into adulthood. The third panel of Figure 7.1 shows the two latent classes characterized by short-term heavy drinking. In one latent class there was heavy drinking during the college-age years only (3%) and in the other there was heavy drinking during young adulthood only (4%). The bottom panel of Figure 7.1 depicts the two latent classes that are characterized by a decreasing trend. One latent class is characterized by heavy drinking in high school and the college-age years, after which heavy drinking ceased (4%). The other latent class is characterized by heavy drinking in high school, the college-age years, and young adulthood; heavy drinking ceased by adulthood (6%).

   As Lanza and Collins (2006) pointed out, it is interesting to note latent classes that did not emerge in this analysis. For example, the model does not include a latent class characterized by heavy drinking during adulthood only, without heavy drinking going back at least as far as young adulthood. Put another way, there was no evidence for heavy drinking beginning during adulthood.

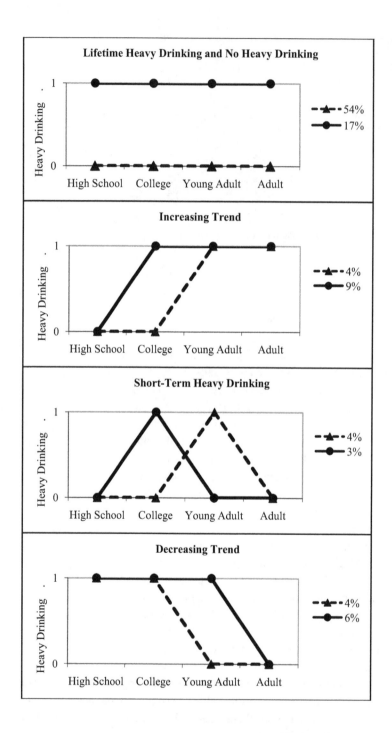

Figure 7.1   Patterns of heavy drinking across four developmental periods, corresponding to the latent classes in Table 7.2 (from Lanza and Collins, 2006). Note that more precise values for item-response probabilities appear in Table 7.1.

Table 7.2   Patterns of Heavy Drinking over Time Corresponding to Eight Latent
Classes (NLSY; N = 1,265) (from Lanza and Collins, 2006)

Label	Prevalence	Developmental Period			
		High School	College Age	Young Adulthood	Adulthood
		Pattern of Heavy Drinking			
No Heavy Drinking	.54	No	No	No	No
High School and College Age	.04	Yes	Yes	No	No
High School, College Age, and Young Adulthood	.06	Yes	Yes	Yes	No
College Age Only	.03	No	Yes	No	No
College Age, Young Adulthood, and Adulthood	.09	No	Yes	Yes	Yes
Young Adulthood Only	.04	No	No	Yes	No
Young Adulthood and Adulthood	.04	No	No	Yes	Yes
Lifetime Heavy Drinking	.17	Yes	Yes	Yes	Yes

7.2.1 Adding a grouping variable

During the college age developmental period, some individuals are enrolled in college
and some are not. Because of the concern in the United States about heavy drinking
among college students, Lanza and Collins (2006) were interested in examining
whether the latent class prevalences were different for those who were in enrolled
in college during the college-age years and those who were not enrolled in college
during that time. This was done by adding a grouping variable to the latent class
model. In the sample used for these analyses, 35 percent were enrolled in college
during the college-age developmental period.

Table 7.3 shows the latent class prevalences for those enrolled and those not
enrolled in college during the college-age years. The most striking difference is that
membership in the College Age Only latent class was estimated to have a prevalence
of 8 percent in the enrolled group, but was extremely rare among the group not
enrolled in college. Those not enrolled in college did engage in heavy drinking
during this developmental period, but it was much more likely to be part of a trend
that extended into other earlier and/or later developmental periods rather than a
short-term phenomenon.

Lanza and Collins next examined the prevalence of heavy drinking at each developmental period by summing across the latent classes. For example, the prevalence
of heavy drinking in high school was obtained by summing the prevalences of the
following latent classes: High School and College Age; High School, College Age,
and Young Adulthood; and Lifetime Heavy Drinking. These prevalences are shown
in Table 7.4. The overall prevalence of heavy drinking for those enrolled in college
and those not enrolled in college was about the same during high school and the
college-age years. However, during young adulthood and adulthood the prevalence

Table 7.3   Prevalences of Latent Classes Representing Patterns of Heavy Drinking over Time for Those Enrolled in College and Those Not Enrolled in College (NLSY; $N = 1,265$) (from Lanza and Collins, 2006)

	Enrolled	Not Enrolled
No Heavy Drinking	.56	.51
High School and College Age	.03	.05
High School, College Age, and Young Adulthood	.09	.06
College Age Only	.08	.00
College Age, Young Adulthood, and Adulthood	.05	.09
Young Adulthood Only	.02	.05
Young Adulthood and Adulthood	.02	.05
Lifetime Heavy Drinking	.15	.18

Table 7.4   Prevalence of Heavy Drinking at Each Developmental Period for Those Enrolled in College and Those Not Enrolled in College (NLSY; $N = 1,265$) (from Lanza and Collins, 2006)

Developmental Period	Enrolled	Not Enrolled
High school	.27	.29
College age	.40	.39
Young adulthood	.33	.43
Adulthood	.22	.33

of heavy drinking was greater for those who had not been enrolled in college during the college-age years.

The example of RMLCA above is a relatively simple one. However, a RMLCA may have any of the features that can be incorporated into a standard LCA. That is, the observed variables may have more than two categories; there may be multiple indicators at each occasion of measurement; and covariates may be incorporated along with a grouping variable.

7.2.2   RMLCA and growth mixture modeling

In this section we have demonstrated how RMLCA can be used to fit models of change over time in longitudinal data. In some ways this approach is similar to growth mixture modeling (e.g., Muthén and Shedden, 1999; Nagin, 2005) in the sense that each latent class is associated with a characteristic vector of responses over time. One important difference between the RMLCA approach to modeling longitudinal data and the growth mixture modeling approach is that whereas in the latent growth curve approach trajectories are fit to express change as a function of time, in RMLCA no functional form of a growth curve is fit. Growth is not characterized as linear, quadratic, cubic, and so on, and no growth parameters per se are estimated. Instead, change over time is modeled in whatever form it naturally occurs in each latent class. This can be helpful when growth is smooth in some latent classes and characterized by a lot of up-and-down or discontinuous change in others.

7.3 LTA

LTA represents a different approach to modeling longitudinal data than the RMLCA approach reviewed in Section 7.2. As will be seen in the remainder of this chapter, rather than modeling an entire vector of times simultaneously for an individual, the emphasis in LTA is on estimating the incidence of transitions from one time to the next. RMLCA and LTA are compared and contrasted in Section 7.13.

LTA is a type of latent Markov model (Everitt, 2006). Early work in this area includes Collins and Wugalter (1992), Langeheine (1988, 1994), Langeheine and van de Pol (1990), van de Pol and de Leeuw (1986), and van de Pol and Langeheine (1990).

7.3.1 Empirical example:
Adolescent delinquency

To introduce LTA, let us revisit and take a closer look at an empirical example that was presented in Chapter 1. Recall that this empirical example is based on $N = 2,087$ adolescents who participated in the Add Health study (Udry, 2003) and were part of the public-use data set. The adolescents were in Grades 10 and 11 (mean age = 16.4) in the 1994–1995 academic year (Wave I), and in Grades 11 and 12 in the 1995–1996 academic year (Wave II). Among the data collected on these adolescents were responses to a series of questionnaire items about delinquent behavior. Specifically, the students were asked how often they had engaged in the following behaviors in the past year: lied to their parents about where or with whom they were; acted loud, rowdy, or unruly in public; damaged property; stolen something from a store; stolen something worth less than $50; and taken part in a group fight. Responses were recoded to "No," meaning the student had not engaged in the behavior in the past year, or "Yes," meaning the student had engaged in the behavior one or more times in the past year.

The proportions of recoded "Yes" responses to each questionnaire item at each wave of data collection are shown in Table 7.5. As discussed in Chapter 1, these marginal proportions of endorsement reveal which of the delinquent behaviors are more normative, such as lying to parents about whereabouts and companions, and which are less normative, such as stealing and fighting. Table 7.5 shows that overall, the proportion of respondents endorsing each item appears to be decreasing over time. The largest decreases are for Publicly loud/rowdy/unruly (from .49 to .40), Lied to parents about where/whom with (from .57 to .49), and Stolen something from store (from .24 to .17). However, these figures provide only a very rough idea of change

Table 7.5 Marginal Response Proportions for Past-Year Delinquency Questionnaire Items (Add Health Public-Use Data, Waves I and II; $N = 2,087$)

	Wave 1		Wave 2	
Item	Observed N	Proportion Responding Yes	Observed N	Proportion Responding Yes
Lied to parents about where/with whom	2,081	.57	1,817	.49
Publicly loud/ rowdy/unruly	2,085	.49	1,817	.40
Damaged property	2,084	.17	1,815	.12
Stolen something from store	2,081	.24	1,816	.17
Stolen something worth < $50	2,085	.20	1,819	.14
Taken part in group fight	2,087	.19	1,819	.16

over time in individual delinquent acts, and provide no information about change over time in broader patterns of delinquent behavior. A look at the N for each item shows that there has been some attrition over time, so the marginals at Wave I are based on a somewhat different sample from those at Wave II, making it difficult to make conclusions about rates of delinquent acts over time by simple inspection of the data.

7.3.2 Why conduct LTA on the adolescent delinquency data?

In previous chapters we have discussed at length how to fit latent class models in order to identify distinct subgroups in data. The latent class models fit to the delinquency data and reported in previous chapters addressed the questions: Are there distinct subgroups of adolescents within the data set that engage in particular patterns of delinquent behavior? If so, what is the distribution of adolescents across these subgroups; in other words, what are the subgroup prevalences?

One reason for conducting LTA on the adolescent delinquency data closely resembles the reason for ever conducting LCA. We consider conducting LTA on this data set because we need a way to represent the complex array of data in Table 7.5 in a format that is more parsimonious and easier to comprehend, and at the same time reveals important scientific information contained in the data. In the adolescent delinquency data the six questionnaire items in Table 7.5 were administered at two times, so the contingency table formed by cross-tabulating the items is very large; it has $W = 2^{12} = 4,096$ cells. Like LCA, LTA can be used to fit a model that represents this large contingency table in a concise way while revealing interesting scientific information.

Another reason for conducting a LTA on the adolescent delinquency data is more conceptual. LCA enables the investigator to address scientific questions about the number and nature of latent classes. When longitudinal data are available, LTA enables the investigator to address an additional set of questions: Is there change between latent classes across time? If so, how can this change be characterized? If an individual is in a particular latent class at Time t, what is the probability that the individual will be in that latent class at Time $t + 1$, and what is the probability that the individual will be in a different latent class? LTA is a way of fitting models that address this set of questions in addition to the set of questions addressed by LCA. Thus, we apply LTA to the adolescent delinquency data because we are interested in studying not only latent classes of adolescent delinquency, but also how individuals transition between latent classes of delinquency over time. Like LCA, LTA estimates item-response probabilities. Therefore, the latent class prevalences and the incidence of transitions between latent classes are estimated while adjusting for measurement error.

In our work in general, and in this book, latent classes in LTA are referred to as *latent statuses*. We prefer to use the term *status* instead of *class* to convey the idea that in LTA the latent classes may be temporary states, and that individuals may move in and out of these states.

7.3.3 Estimation and assessing model fit

Just as in LCA, the starting point for LTA is a contingency table formed by cross-tabulating all the observed variables. In LTA these contingency tables are often very large, because each variable is measured at two or more times. In LCA of the delinquency example described in Chapter 1, there were six questionnaire items measured at one time, so the contingency table had $W = 2^6 = 64$ cells. As mentioned above, in the LTA the same questionnaire items were measured at two times, so the contingency table now has $W = 2^{12} = 4,096$ cells. Adding one or more times expands the size of the contingency table exponentially. The objective of fitting LTA models is to identify a model that represents this contingency table well.

Latent transition models tend to have very large degrees of freedom, because W tends to be very large. However, identification problems can occur in LTA, even in models with thousands of degrees of freedom. The primary reason for this is that the large contingency tables to which LTA is usually applied tend to be sparse (see Chapter 4). Imposing parameter restrictions so that measurement of the latent statuses is the same at each measurement occasion helps tremendously with identification problems; this is discussed in more detail in Section 7.10.

The combination of very large degrees of freedom and extreme sparseness renders traditional hypothesis testing of the absolute fit of a latent transition model nearly useless. This is because the distribution of the G^2 is not well represented by the chi-square distribution, and therefore p-values are extremely inaccurate. Therefore, we prefer to frame model selection in relative terms wherever possible, that is, to fit a series of models and rely on the AIC and BIC to make decisions about which appears to represent the data best. Hypothesis testing can be useful when comparing nested latent transition models. Of course, parsimony and conceptual appeal are important model selection criteria in LTA, just as they are in LCA.

Sometimes fitting a series of latent class models with different numbers of latent classes to data at a single time point can be helpful as a preliminary step in model selection in LTA. Doing this at each time point can be informative about the latent structure within times, and how that structure changes across time points. However, this should be viewed as a preliminary step, not as a definitive answer about the number of latent statuses in the latent transition model. It is important to note that the best-fitting latent class model at any given time may not correspond to the best-fitting latent transition model fit to all occasions of measurement. The best-fitting model based on the data from all occasions of measurement may include a different number of latent statuses than the number of latent classes identified at one particular time. One reason is related to statistical power; the additional information provided by multiple occasions of measurement sometimes enables detection of additional latent classes. Another reason is that LCA performed on data from individual measurement occasions may overlook some latent statuses on some occasions, particularly if the longitudinal data span more than one developmental period. In LTA some latent statuses may have very low prevalences at earlier occasions but as individuals transition into them over time, their prevalences increase. Other latent statuses may be highly prevalent on earlier occasions but their prevalences may decline later when individuals have transitioned out of them. In either case, if a latent class model is fit to data from an occasion in which the prevalence of a latent status is low, it may not be detected as a latent class. In addition, differential rates of missing data across times may limit the usefulness of examining the latent class structure at one particular time. For these reasons we recommend conducting LCA at one or more time points as a preliminary step to inform model selection in LTA, but that selection of the LTA model be based primarily on an assessment of fit using data from all occasions of measurement.

7.3.4 Model fit in the adolescent delinquency example

Table 7.6 summarizes the fit statistics associated with a variety of LTA models fit to the delinquency data. In the next section we discuss the details of how the degrees of freedom were computed; for now we note that just as in LCA, the degrees of freedom associated with a LTA model equal $W - P - 1$, where P is the number of parameters estimated in the model. As Table 7.6 shows, models ranging from two to six latent statuses were fit. Table 7.6 contains degrees of freedom and G^2 values, but it does not include the p-values associated with the G^2's. Ordinarily, p-values are determined

Table 7.6   Summary of Information for Selecting Number of Latent Statuses of Adolescent Delinquency at Two Times (Add Health Public-Use Data, Waves I and II, $N = 2,087$)

Number of Latent Statuses	Number of Parameters Estimated	G^{2*}	df	AIC	BIC	ℓ
2	15	2,713.1	4,080	2,743.1	2,827.8	−11,034.4
3	26	1,888.6	4,069	1,940.6	2,087.4	−10,622.2
4	39	1,714.5	4,056	1,792.5	2,012.6	−10,535.1
5	54	1,588.0	4,041	1,696.0	2,000.7	−10,471.8
6	71	1,514.7	4,024	1,656.7	2,057.4	−10,435.2

$^*p$-values not reported because the degrees of freedom are too large.

by referring to the χ^2 distribution. In this example $N/W = 2,087/4,096 = .51$, suggesting that the distribution of the G^2 will not be well enough approximated by the χ^2 distribution for p-values to be accurate. Instead, we rely primarily on the AIC and BIC for model selection.

Figure 7.2 shows G^2, AIC, and BIC for the latent transition models that were fit to the delinquency data as a function of the number of latent statuses fit in each model.

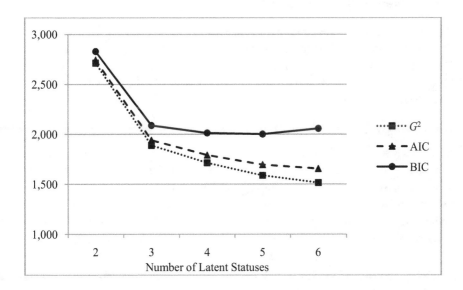

Figure 7.2 G^2, AIC, and BIC for latent transition models of adolescent delinquency (Add Health public-use data, Waves I and II; $N = 2,087$).

The figure shows that the AIC and BIC both declined through five latent statuses. At six latent statuses the AIC continued to decline, but the BIC rose. Thus the AIC points to the six-latent-status model, and the BIC points to the five-latent-status model. The five-latent-status model was conceptually appealing and also more parsimonious, so we selected this one. (More information about model selection may be found in Chapter 4.) This is the model that was reported in Table 1.4 and is repeated here for convenience as Table 7.7. When these findings were presented in Chapter 1, the objective was to provide an initial feel for LTA. Here we revisit this familiar example in more specific and technical terms. In Sections 7.4 and 7.5 we use the delinquency data to illustrate the LTA model and notation.

7.4   LTA MODEL PARAMETERS

Three different sets of parameters are estimated in LTA, two of which have direct counterparts in LCA.

7.4.1   Latent status prevalences

Recall from Chapter 1 the labels we assigned to the five latent statuses of adolescent delinquency: Nondelinquents, Liars, Verbal Antagonists, Shoplifters, and General Delinquents. The prevalences of these latent statuses appear in the first section of Table 7.7. Latent status prevalences in LTA play essentially the same role that latent class prevalences play in LCA. The main difference between latent class prevalences and latent status prevalences is that there is a vector of latent status prevalences at each time. As noted in Chapter 1, at Time 1 the Liars latent status was the most prevalent, followed closely by the Nondelinquents and Verbal Antagonists latent statuses, which were about equally prevalent. Next is Shoplifters and General Delinquents, which were the least prevalent latent statuses. At Time 2 the overall pattern is similar, except that Nondelinquents was the most prevalent latent status, followed by Liars and Verbal Antagonists. The Shoplifters and General Delinquents latent statuses remained the least prevalent. There appears to be a slight decrease in delinquency over time; the Verbal Antagonists, Shoplifters, and General Delinquents latent status prevalences declined between Times 1 and 2, and the Nondelinquents latent status prevalence increased.

However, based on the latent status prevalences at each time alone, it is impossible to tell how and to what extent individuals were moving between latent statuses. As we discuss in Section 7.4.3, this information is contained in a different set of parameters, the transition probabilities.

Table 7.7   Five-Latent-Status Model of Past-Year Delinquency (Add Health
Public-Use Data, Waves I and II; $N = 2{,}087$; Table 1.4 repeated for convenience)

	Latent Status				
	Non-delinquents	Liars	Verbal Antagonists	Shoplifters	General Delinquents
Latent status prevalences					
Time 1 (grades 10/11)	.24	.27	.25	.13	.10
Time 2 (grades 11/12)	.34	.28	.22	.10	.06
*Item-response probabilities of a Yes response**					
Lied to parents	.00	.72[†]	.72	.74	**.89**
Publicly loud/rowdy/unruly	.15	.22	**.89**	.48	**.92**
Damaged property	.00	.05	.25	.17	**.68**
Stolen something from store	.02	.02	.04	**.92**	**.90**
Stolen something worth $< \$50$	.00	.00	.06	**.72**	**.85**
Taken part in group fight	.03	.07	.34	.17	**.54**
Probability of transitioning to... Conditional on... ...Time 1 latent status	*...Time 2 latent status*				
Nondelinquents	**.84**[‡]	.10	.03	.02	.01
Liars	.22	**.67**	.00	.10	.00
Verbal Antagonists	.14	.13	**.65**	.04	.04
Shoplifters	.26	.29	.11	**.34**	.00
General Delinquents	.08	.01	.36	.09	**.45**

$\ell = -10{,}471.8$. *Item-response probabilities constrained equal across times.

[†] Item-response probabilities $> .5$ in bold to facilitate interpretation.

[‡] Diagonal transition probabilities in bold to facilitate interpretation.

7.4.2   Item-response probabilities

The second section of Table 7.7 contains the item-response probabilities corresponding to a "Yes" response to each delinquency item. (Because the item-response probabilities corresponding to the "No" responses are the complements of those corresponding to the "Yes" responses, we do not report them here.) The item-response probabilities in LTA play the same role that they play in LCA; that is, they form the basis for assigning labels to latent statuses in LTA. Moreover, considerations of homogeneity and latent class separation apply to the latent statuses in LTA in exactly the same way as they apply to the latent classes in LCA. An important difference between LCA and LTA is that in LTA, there is a set of item-response probabilities corresponding to each time. In this example, there are two sets of item-response probabilities, one for Time 1 and one for Time 2. However, parameter restrictions have been imposed so that the item-response probabilities are equal across the two times. Therefore, only a single set is reported here. (In Section 7.11 we use a different empirical example to demonstrate how to test the hypothesis that the item-response probabilities in a LTA are equal across times.) This type of restriction on the item-response probabilities is commonly used in LTA. It helps with model identification and, importantly, ensures that the meaning of the latent statuses remains constant over time.

The larger conditional probabilities in Table 7.7 appear in bold font to highlight the overall pattern. The Nondelinquents latent status is characterized by very low probabilities of responding "Yes" to any of the items. Those in the Liars latent status had a high probability of reporting lying to parents; those in the Verbal Antagonists latent status had a high probability both of reporting lying and of reporting being publicly loud. The Shoplifters latent status is characterized by high probabilities of reporting having lied to parents, stolen something from a store, and stolen something worth less than $50. Finally, those in the General Delinquents latent status had a high probability of endorsing all of the questionnaire items.

This pattern of item-response probabilities shows good overall homogeneity and latent class separation. (See Chapter 3 for a thorough discussion of these terms.) Recall that when latent class c is highly homogeneous, members of c are likely to provide the same observed response pattern, and that one response pattern is highly characteristic of (i.e., highly probable for) latent class c. Most of the latent statuses in Table 7.7 are characterized by one response pattern. For example, the Nondelinquents latent status is characterized by the response pattern (No, No, No, No, No, No). The Shoplifters and General Delinquents latent statuses are somewhat less homogeneous than the others. Members of the Shoplifters latent status were about equally likely to respond "No" (.52) or "Yes" (.48) to the questionnaire item concerning public loudness. Thus this latent status is characterized by two different response patterns: (Yes, No, No, Yes, Yes, No) and (Yes, Yes, No, Yes, Yes, No), with the former slightly more probable given membership in the Shoplifters latent status than the latter. Members of the General Delinquents latent status were about equally likely to respond "No" (.46) or "Yes" (.54) to the questionnaire item about taking part in a group fight. Thus this latent status is also characterized by two different response patterns. The more probable response pattern is one made up entirely of "Yes" responses, but only slightly less probable is a response pattern in which all of the responses are "Yes" except the response to the last questionnaire item. It is informative, however, that General Delinquents have a substantially higher probability (.54) of reporting having taken part in a group fight than do individuals in the other four latent statuses (.03 to .34).

Recall that when there is a high degree of latent class separation, a response pattern that emerges as characteristic of (i.e., highly probable for) a particular latent class will be characteristic of that latent class only and will not be characteristic of any of the other latent classes. Latent class separation is evident in the item-response probabilities in Table 7.7; each latent status is distinct from others. For example, only one latent status, Verbal Antagonists, is characterized by a response pattern consisting of "Yes" responses to the first two items and "No" responses to the remaining items.

7.4.3   Transition probabilities

The third set of parameters estimated in LTA has no direct counterpart in LCA. These are the transition probabilities, which are shown in the third section of Table 7.7. The transition probabilities are often of primary interest, because they express how change occurs between latent statuses over time. Transition probabilities are typically arranged in a matrix with the rows corresponding to the earlier time and the columns corresponding to the later time. There are $T - 1$ transition probability matrices; in this example $T = 2$, so there is one transition probability matrix. The transition probabilities express the incidence of transitioning to the column latent status, conditional on earlier membership in the row latent status. The diagonal elements of the transition probability matrix represent the probability of being in a particular latent status at one time conditional on being in that same latent status at the previous time. In this example, this can be considered the probability of remaining stable in delinquency over time. To assist in interpretation, the diagonal elements of the transition probability matrix appear in bold in the table.

The transition probability matrix shows that those who were in the Nondelinquents latent status at Time 1 had a high probability (.84) of remaining there for the following year. If they did transition, they were most likely to transition to the Liars latent status (.10), which represents a relatively mild form of acting out. Those in the Liars latent status at Time 1 had a .67 probability of being there at Time 2; their most likely transition was back to the Nondelinquents latent status (.22). Those in the Verbal Antagonists latent status at Time 1 had a .65 probability of remaining there. They were less likely to transition to the Shoplifters or General Delinquents latent status (.04 for each) than they were to transition to the Nondelinquents or Liars latent status (.14 and .13, respectively). Those in the Shoplifters and General Delinquents latent statuses were least likely to remain in the same latent status (.34 and .45, respectively). Shoplifters had a negligible (rounded to .00) probability of transitioning to the more serious latent status of General Delinquents. They were most likely to transition to the Nondelinquents (.26) or Liars (.29) latent statuses. General Delinquents were most likely to transition to the Verbal Antagonists (.36) latent status. Only 8 percent of those who were General Delinquents at Time 1 were expected to be Nondelinquents at Time 2, compared to 14 percent of Verbal Antagonists, 22 percent of Liars, and 26 percent of Shoplifters.

Taken as a whole, the information about change represented by the matrix of transition probabilities presents a parsimonious yet detailed picture of how adolescents move into and out of delinquency latent statuses during high school. Note that because each row of the transition probability matrix is conditioned on Time 1 latent status, this analysis of change in delinquency over time implicitly controls for Time 1 delinquency.

7.5 LTA: MODEL AND NOTATION

In this section the latent transition model is introduced in a more formal and general way. To simplify the exposition, this presentation of the latent transition model assumes no missing data on the observed indicator variables. However, missing data on observed variables is handled in LTA the same way that it is handled in LCA (see Chapter 4). Data can be used from individuals who provide responses to only some questions at a particular time and from those who are present at only a subset of the times of measurement. The usual missing at random (MAR) assumption, discussed in Section 4.2.3, is made.

Suppose that there are $j = 1, ..., J$ observed indicator variables, and the J observed variables have been measured at $t = 1, ..., T$ times. In the adolescent delinquency example, the observed variables are the delinquency questionnaire items, so $J = 6$. Observed variable j has $r_{j,t} = 1, ..., R_{j,t}$ response categories. To keep the exposition simple we will assume that in general the number of response categories for each observed indicator variable is identical across times; that is, $R_{j,1} = R_{j,2} = \cdots = R_{j,T} = R_j$ for all j. Each of the questionnaire items in the adolescent delinquency example has two response alternatives, "No" and "Yes," so in this example $R_j = 2$ for all j. The contingency table formed by cross-tabulating the J variables at T times has $W = \prod_{t=1}^{T} \prod_{j=1}^{J} R_j$ cells. The questionnaire was administered at two times, so $T = 2$. Thus $W = 2^{2 \times 6} = 2^{12} = 4,096$.

Corresponding to each of the W cells in the contingency table is a complete response pattern. In LTA this is a vector of responses to the J variables at each of the T times. The response patterns are represented by $\mathbf{y} = (r_{1,1}, ..., r_{J,T})$. For instance, in the delinquency example a response pattern of (Yes, No, No, No, No, No, Yes, Yes, No, No, No, No) represents responses of "Yes" to the first questionnaire item, Lied to parents about where/with whom, at Times 1 and 2, a response of "Yes" to the second questionnaire item, Publicly loud/rowdy/unruly at Time 2 only, and responses of "No" everywhere else. Let $\mathbf{Y}$ refer to the array of response patterns. $\mathbf{Y}$ has W rows and $T \times J$ columns. Each response pattern $\mathbf{y}$ is associated with probability $P(\mathbf{Y} = \mathbf{y})$, and $\sum P(\mathbf{Y} = \mathbf{y}) = 1$.

Let L represent the categorical latent variable overall. L has S latent statuses. Let L_1 represent the categorical latent variable at Time 1, where $s_1 = 1, ..., S$; let L_2 represent the categorical latent variable at Time 2, where $s_2 = 1, ..., S$ at Time 2, and so on, up to L_T, representing the categorical latent variable at Time T, with $s_T = 1, ..., S$. For simplicity we assume for purposes of this discussion that the number of latent statuses is identical across times (i.e., $S_1 = S_2 = \cdots = S_T = S$). However, it is technically possible for a latent status to be empty at a particular time.

As discussed above, three different sets of parameters are estimated in LTA.

7.5.0.1 *Latent status prevalences*

δ_{s_t} represents the prevalence of latent status s at Time t, or in other words, the probability of membership in latent status s at Time t. In the delinquency example there are five latent statuses at each of two times, for a total of 10 δ's. For instance, $\delta_{2_1} = .27$ (see Table 7.7). This is the probability of membership in the Liars latent status at Time 1.

Because the latent statuses are mutually exclusive and exhaustive at each time, that is, each individual is a member of one and only one latent status at Time t,

$$\sum_{s_t=1}^{S} \delta_{s_t} = 1. \tag{7.1}$$

In other words, within a particular Time t the latent status prevalences sum to 1. Examination of Table 7.7 confirms this, within rounding error.

7.5.0.2 *Item-response probabilities*

$\rho_{j,r_{j,t}|s_t}$ represents the probability of response $r_{j,t}$ to observed variable j, conditional on membership in latent status s_t at Time t. As mentioned above, in the model shown in Table 7.7 the item-response probabilities were constrained to be equal across the two times, so only one set of parameters is shown. For example, assuming that a response of "Yes" is designated with a 1, $\rho_{3,1|5_1} = \rho_{3,1|5_2} = .68$ represents the probability of endorsing the item Damaged property, conditional on membership in the General Delinquents latent status at both Times 1 and 2.

For each combination of latent status s, observed variable j, and Time t, there are R_j item-response probabilities. Because each individual provides one and only one response alternative to variable j at a particular time t,

$$\sum_{r_{j,t}=1}^{R_j} \rho_{j,r_{j,t}|s_t} = 1 \tag{7.2}$$

for all j, t. In other words, for individuals in latent status s_t at Time t, the probabilities of each of the response alternatives to variable j sum to 1.

7.5.0.3 *Transition probabilities*

$\tau_{s_{t+1}|s_t}$ represents the probability of a transition to latent status s at Time $t+1$, conditional on membership in latent status s at Time t. For example, $\tau_{2_2|4_1} = .29$ represents the probability of transitioning to the Liars latent status at Time 2, conditional on membership in the Shoplifters latent status at Time 1.

As shown in Table 7.7, the τ's are often arranged in a transition probability matrix as follows:

$$\begin{bmatrix} \tau_{1_{t+1}|1_t} & \tau_{2_{t+1}|1_t} & \cdots & \tau_{S_{t+1}|1_t} \\ \tau_{1_{t+1}|2_t} & \tau_{2_{t+1}|2_t} & \cdots & \tau_{S_{t+1}|2_t} \\ \cdots & \cdots & \cdots & \cdots \\ \tau_{1_{t+1}|S_t} & \tau_{2_{t+1}|S_t} & \cdots & \tau_{S_{t+1}|S_t} \end{bmatrix}. \quad (7.3)$$

Among the individuals who are in latent status s_t at Time t, each individual is in one and only one latent status at Time $t+1$, s_{t+1}. s_{t+1} may represent the same latent status as s_t or it may be a different latent status. At each , latent status membership is mutually exclusive and exhaustive; that is, individuals belong to one and only one latent status at each time. Therefore,

$$\sum_{s_{t+1}=1}^{S} \tau_{s_{t+1}|s_t} = 1. \quad (7.4)$$

In other words, each row of the transition probability matrix sums to 1. This is evident in Table 7.7, within rounding error.

7.5.1 Fundamental expression

Now we examine a mathematical expression that is fundamental to latent transition models. Let us establish an indicator function $I(y_{j,t} = r_{j,t})$ that equals 1 when the response to variable $j = r_j$ at Time t, and equals 0 otherwise. (As has been mentioned in previous chapters, this function is merely a device for picking out the appropriate ρ parameters to multiply together.) Then Equation 7.5 expresses how the probability of observing a particular vector of responses is a function of the probabilities of membership in each latent status at Time 1 (the δ_{s_1}'s), the probabilities of transitioning to a latent status at a particular time conditional on latent status membership at the immediately previous time (the τ's), and the probabilities of observing each response at each time conditional on latent status membership (the ρ's):

$$P(\mathbf{Y} = \mathbf{y}) = \sum_{s_1=1}^{S} \cdots \sum_{s_T=1}^{S} \delta_{s_1} \tau_{s_2|s_1} \cdots \tau_{s_T|s_{T-1}} \prod_{t=1}^{T} \prod_{j=1}^{J} \prod_{r_{j,t}=1}^{R_j} \rho_{j,r_{j,t}|s_t}^{I(y_{j,t}=r_{j,t})}. \quad (7.5)$$

If there are two times, Equation 7.5 reduces to

$$P(\mathbf{Y} = \mathbf{y}) = \sum_{s_1=1}^{S} \sum_{s_2=1}^{S} \delta_{s_1} \tau_{s_2|s_1} \prod_{t=1}^{2} \prod_{j=1}^{J} \prod_{r_{j,t}=1}^{R_j} \rho_{j,r_{j,t}|s_t}^{I(y_{j,t}=r_{j,t})}. \quad (7.6)$$

7.6   DEGREES OF FREEDOM ASSOCIATED
WITH LATENT TRANSITION MODELS

The degrees of freedom in LTA are computed in much the same way as in LCA. The starting point is the number of cells in the contingency table, W. Then $df = W - P_\delta - P_\rho - P_\tau - 1$, where P_δ represents the number of latent status prevalences estimated; P_ρ represents the number of item-response probabilities estimated; and P_τ represents the number of transition probabilities estimated. Computation of P_δ, P_ρ, and P_τ when there are no parameter restrictions is discussed in this section. Imposing parameter restrictions reduces the number of parameters estimated. This is discussed in Section 7.10.

7.6.1   Computing the number of
latent status prevalences
estimated

The tables in this book that show parameter estimates, for example, Table 7.7, include the latent status prevalences for all T times. However, it is important to keep in mind that in the latent transition models discussed in this book, *only the latent status prevalences associated with Time 1 are estimated independently*. The latent status prevalences for Times 2 through T are computed based on the latent status prevalences for Time 1 and the transition probabilities, as follows:

$$\delta_{s_t} = \sum_{s_{t-1}=1}^{S} \delta_{s_{t-1}} \tau_{s_t|s_{t-1}} \qquad (7.7)$$

for all $t \geq 2$.

In other words, an individual who is a member of latent status s at Time t could have been in any of the latent statuses at the immediately previous time, Time $t - 1$, and transitioned from that latent status into latent status s. Thus the probability of membership in latent status s at Time t is a function of the probabilities of membership in each latent status at Time $t - 1$ and the conditional probability of transitioning from each latent status into latent status s between Time $t - 1$ and Time t.

For example, once the latent status prevalences at Time 1 and the transition probabilities between Times 1 and 2 have been estimated, the Time 2 latent status prevalences can be computed based on these estimates. Using the results in Table 7.7, the probability of membership in the Verbal Antagonists latent status at Time 2 can be computed using Equation 7.7:

$$
\begin{aligned}
\delta_{3_2} &= (\delta_{1_1} \times \tau_{3_2|1_1}) + (\delta_{2_1} \times \tau_{3_2|2_1}) + (\delta_{3_1} \times \tau_{3_2|3_1}) + (\delta_{4_1} \times \tau_{3_2|4_1}) \\
&\quad + (\delta_{5_1} \times \tau_{3_2|5_1}) \\
&= (.24 \times .03) + (.27 \times .00) + (.25 \times .65) + (.13 \times .11) + (.10 \times .36) \\
&= .22 \tag{7.8}
\end{aligned}
$$

(within rounding error).

In a model that involves more than two times, the latent status prevalences for Times 3 through T can be computed by first computing the latent status prevalences for Time 2 as demonstrated in Equation 7.8, then applying Equation 7.7 to compute the latent status prevalences for Time 3 using the Time 2 latent status prevalences and the probabilities of transitions between Times 2 and 3, and so on.

Equation 7.1 shows that one latent status prevalence can always be obtained by subtraction. Thus, if no parameter restrictions have been imposed,

$$
P_\delta = S - 1. \tag{7.9}
$$

In the delinquency example, $P_\delta = 5 - 1 = 4$.

7.6.2 Computing the number of item-response probabilities estimated

Equation 7.2 shows that for each combination of latent status s, observed variable j, and Time t, the R_j item-response probabilities sum to 1; thus one item-response probability can be obtained by subtraction. Therefore, if no parameter restrictions have been imposed,

$$
P_\rho = ST \sum_{j=1}^{J} (R_j - 1). \tag{7.10}
$$

In the delinquency example, there would have been $P_\rho = 5 \times 2 \times (1+1+1+1+1+1) = 5 \times 2 \times 6 = 60$ item-response probabilities estimated in an unrestricted model. However, in this example all item-response probabilities were constrained to be equal over time. Therefore, there were only $P_\rho = 5 \times 1 \times (1+1+1+1+1+1) = 5 \times 6 = 30$ item-response probabilities estimated in the model shown in Table 7.7.

7.6.3 Computing the number of transition probabilities estimated

Because by Equation 7.4 each row of the transition probability matrix sums to 1, it follows that one transition probability in each row can be obtained by subtraction.

Therefore, in each transition probability matrix at most $S(S - 1)$ parameters are estimated. There are $T - 1$ transition probability matrices. Thus if no parameter restrictions are imposed,

$$P_\tau = (T - 1)S(S - 1). \tag{7.11}$$

In the delinquency example, $P_\tau = 1 \times 5 \times 4 = 20$.

7.7 EMPIRICAL EXAMPLE: ADOLESCENT DEPRESSION

This empirical example is drawn from a previous book chapter of ours: Lanza, Flaherty, and Collins (2003). The example is based on the Add Health study (Udry, 2003) public-use data set, but the sample consists of a slightly different subset of participants than the adolescent delinquency example. This subset of participants consists of $N = 2{,}061$ adolescents who were in Grades 10 or 11 at the first wave of data collection (1994–1995) and who provided a response to at least one of the questionnaire items used in this analysis and to all of the covariates to be introduced in Chapter 8. The observed indicator variables are responses to the eight questionnaire items listed in Table 7.8, all of which ask about depressed feelings in the past week. Below we provide the complete wording of each item; the tables in this chapter and the next will contain a shortened version of the wording.

The first four questionnaire items asked about feelings of sadness. These were: You felt that you could not shake off the blues, even with help from your family and your friends; You felt depressed; You felt lonely; and You felt sad. The next two asked about feeling disliked. These were: People were unfriendly to you; and You felt that people disliked you. The last two asked about feelings of failure. These were: You thought your life had been a failure; and You felt life was not worth living.

Table 7.8 Marginal Response Proportions for Adolescent Past-Week Depression Questionnaire Items (Add Health Public-Use Data, Waves I and II; $N = 2{,}061$)

	Time 1		Time 2	
	Grades 10/11		Grades 11/12	
Item	Observed N	Proportion Endorsing Item	Observed N	Proportion Endorsing Item
Couldn't shake blues	2,060	.32	1,805	.32
Felt depressed	2,060	.41	1,806	.40
Felt lonely	2,060	.39	1,806	.37
Felt sad	2,061	.50	1,806	.47
People unfriendly	2,061	.34	1,806	.34
Felt disliked	2,060	.35	1,806	.32
Life was failure	2,061	.16	1,803	.16
Life not worth living	2,061	.11	1,802	.09

Participants reported how often they had had each feeling during the past week. A six-category scale was used, where the response categories were "Never," "Rarely," "Sometimes," "A lot," "Most of the time," and "All of the time." For purposes of these analyses these highly skewed variables were recoded to two categories. An item was coded as endorsed if the respondent reported having experienced the feeling "Sometimes," "A lot," "Most of the time," and "All of the time." If the respondent reported having experienced the feeling "Never" or "Rarely," the item was coded as not endorsed.

Each item's proportion of endorsement at Times 1 and 2 appears in Table 7.8. The most frequently endorsed items at each time are Felt sad and Felt depressed. The two items about failure are considerably less frequently endorsed than the others. There does not appear to be much of a trend in these marginal proportions across the two times, except for a slight tendency for less endorsement of the items at Time 2. However, as the table shows, there has been some attrition between Times 1 and 2. This makes it difficult to interpret this slight trend. Also, the information in Table 7.8 provides no insight into individual-level trends over time in depressive symptoms.

LTA enables us to address some interesting questions using this data set. Are there subgroups characterized by different profiles of symptoms of depression? How do individuals move between subgroups over time? Perhaps the reason why the marginals in Table 7.8 are so similar over time is that individuals' symptoms of depression change very little. Alternatively, perhaps as some people stop experiencing a symptom of depression, others start experiencing the same symptom, so that although there is change, the overall rates of symptom prevalence are maintained at a relatively stable level over time.

Table 7.9 summarizes the results of fitting latent transition models ranging from two to seven latent statuses to the adolescent depression data. Table 7.9 contains degrees of freedom and G^2 values, but it does not include the p-values associated with the G^2's, for the same reason that they were not included in Table 7.6; with $W = 2^{16} = 65,536$ cells in the contingency table, p-values determined by referring to the χ^2 distribution will be inaccurate, so we will rely mainly on the AIC and BIC for model selection. Figure 7.3 shows G^2, AIC, and BIC for the latent transition models that were fit to the depression data as a function of the number of latent statuses. As Table 7.9 and Figure 7.3 show, the AIC declines as each larger model is fit, with the lowest AIC corresponding to the seven-latent-status model. However, the lowest BIC corresponds to the five-latent-status model. We chose the five-latent-status model because we found it to be conceptually appealing, and in addition it is more parsimonious than the model with six latent statuses. Table 7.10 shows the parameter estimates corresponding to the five-latent-status model.

Table 7.9   Summary of Information for Selecting Number of Latent Statuses of Adolescent Depression (Add Health Public-Use Data, Waves I and II; $N = 2,061$)

Number of Latent Statuses	Number of Parameters Estimated	G^{2*}	df	AIC	BIC	ℓ
2	19	6,627.1	65,516	6,665.1	6,772.1	−15,291.7
3	32	5,860.3	65,503	5,924.3	6,104.4	−14,908.3
4	47	5,367.5	65,488	5,461.5	5,726.2	−14,661.9
5	64	5,112.9	65,471	5,240.9	5,601.3	−14,534.7
6	83	4,997.6	65,452	5,163.6	5,631.0	−14,477.0
7	104	4,901.3	65,431	5,109.3	5,694.9	−14,428.8

$^*p$-values not reported because the degrees of freedom are too large.

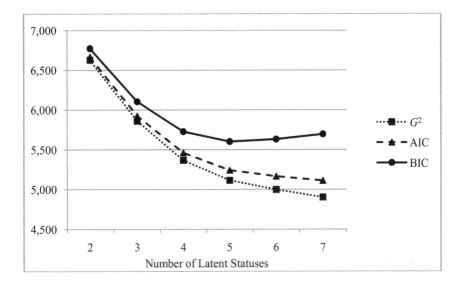

Figure 7.3 G^2, AIC, and BIC for latent transition models of adolescent depression (Add Health public-use data, Waves I and II; $N = 2,061$).

7.7.1   Latent status prevalences

As Table 7.10 shows, we assigned the following labels to the latent statuses: Not Depressed; Sad; Disliked; Sad + Disliked; and Depressed. At each time Not Depressed was the most prevalent latent status (.36 at Time 1 and .40 at Time 2), followed by Sad (.23 at both times). Sad + Disliked and Depressed were the least prevalent. The

Table 7.10 Five-Latent-Status Model of Adolescent Depression (Add Health Public-Use Data, Waves I and II; $N = 2,061$)

	Latent Status				
	Not Depressed	Sad	Disliked	Sad + Disliked	Depressed
Latent status prevalences					
Time 1 (grades 10/11)	.36	.23	.18	.11	.11
Time 2 (grades 11/12)	.40	.23	.15	.11	.10
Item-response probabilities					
*of a Yes response**					
Couldn't shake blues	.03	**.52**[†]	.12	**.69**	**.81**
Felt depressed	.06	**.68**	.18	**.84**	**.94**
Felt lonely	.06	**.55**	.27	**.84**	**.85**
Felt sad	.11	**.80**	.34	**.92**	**.94**
People unfriendly	.09	.16	**.68**	**.76**	**.65**
Felt disliked	.03	.13	**.67**	**.88**	**.79**
Life was failure	.01	.13	.08	.13	**.89**
Life not worth living	.00	.06	.05	.06	**.70**
Probability of transitioning to...		...Time 2 latent status			
Conditional on...					
...Time 1 latent status					
Not Depressed	**.75**[‡]	.12	.09	.02	.02
Sad	.30	**.54**	.04	.07	.04
Disliked	.25	.05	**.47**	.17	.06
Sad + Disliked	.03	.34	.17	**.36**	.11
Depressed	.08	.16	.03	.14	**.59**

$\ell = -14,534.7$.

*Item-response probabilities constrained equal across times.

[†]Item-response probabilities $> .5$ in bold to facilitate interpretation.

[‡]Diagonal transition probabilities in bold to facilitate interpretation.

prevalences of the adolescent depression latent statuses did not change very much across the two times.

7.7.2 Item-response probabilities

The middle section of Table 7.10 contains the item-response probabilities corresponding to endorsement of each item. Larger item-response probabilities are in bold font to aid in interpretation. For this model the item-response probabilities were constrained to be equal across times, so only one set of these parameters is reported here. (In Section 7.11.1 we test the hypothesis that the item-response probabilities are equal across time.) Those in the Not Depressed latent status had a low probability of endorsing any of the depression items. Those in the Sad latent status were likely to endorse the four items asking about sad feelings and were unlikely to endorse the items asking about feeling disliked or having feelings of failure. The Disliked latent status is characterized by reporting that people are unfriendly and feeling disliked, but not reporting feelings of sadness or failure. Those in the Sad + Disliked latent status had a high probability of endorsing each of the items except those concerning feelings of failure, and those in the Depressed latent status had a high probability of endorsing each of the eight items.

This empirical example demonstrates how there can be both qualitative and quantitative differences among latent statuses. Many of the latent statuses in this example can be ordered along a continuum representing increasing severity of depression. For example, the latent statuses Not Depressed, Sad, Sad + Disliked, and Depressed represent increasing depression. Similarly, the latent statuses Not Depressed, Disliked, Sad + Disliked, and Depressed represent increasing depression. However, it is difficult to see how the latent statuses Sad and Disliked can be ordered. They seem to represent different subtypes of depression that do not appear immediately to differ in severity; in other words, the difference between the Sad and Disliked latent statuses is qualitative in nature.

This overall pattern of item-response probabilities shows good homogeneity and latent status separation. The Not Depressed, Disliked, Sad + Disliked, and Depressed latent status are each characterized by a single response pattern. For example, the Not Depressed latent status is characterized by the response pattern consisting of all "No" responses to the items, and the Depressed latent status is clearly characterized by the response pattern consisting of all "Yes" responses to the items. The Sad latent status is less homogeneous than the others. It is most clearly characterized by the response pattern consisting of endorsing the first four items, but is nearly as well characterized by the following response patterns: endorsing You felt that you could not shake off the blues, even with help from your family and your friends, You felt depressed and You felt sad; endorsing You felt depressed, You felt lonely, and You felt sad; and endorsing only You felt depressed and You felt sad. Latent class separation is evident in Table 7.10; in no case is a response pattern that characterizes one of the latent statuses also highly probable in a different latent status. Even though the Sad latent status can be characterized by several different response patterns, none of these response patterns is highly likely for a different latent status.

7.7.3   Transition probabilities

The lowest section of Table 7.10 shows the transition probability matrix, with the diagonal elements in bold font. The elements of the transition probability matrix in Table 7.10 are depicted in Figure 7.4. Although Table 7.10 shows that the latent status prevalences are very similar at Times 1 and 2, and both Table 7.10 and Figure 7.4 show that in general an individual's most likely latent status at Time 2 was his or her latent status membership at Time 1, there is nevertheless considerable movement between latent statuses. For example, those in the Sad latent status at Time 1 have a 30 percent chance of being in the Not Depressed latent status at Time 2. Those in the Disliked latent status have a 17 percent chance of transitioning to the Sad + Disliked

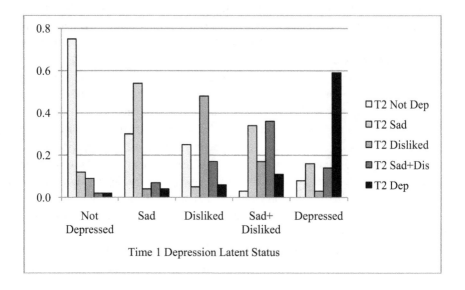

Figure 7.4 Transition probabilities for latent transition model of adolescent depression (Add Health public-use data, Waves I and II; $N = 2,061$).

latent status at Time 2, and a 25 percent chance of transitioning to Not Depressed. Those in Sad + Disliked at Time 1 have a 34 percent chance of being in the Sad latent status and a 17 percent chance of being in the Disliked latent status at Time 2. Finally, those in the Depressed latent status have a 14 percent chance of being in the Sad + Disliked latent status and a 16 percent chance of being in the Sad latent status at Time 2. It is impossible to determine how much movement between latent statuses actually took place between measurement occasions, because according to this model many transitions may have occurred between when adolescents were asked about past-week depression symptoms at Time 1 and when they were asked again a year later. Thus, one way to interpret the transition probabilities in this example is as a lower bound on off-diagonal elements (i.e., we may be underestimating change over time) and an upper bound on diagonal elements. (We return to the issue of how to interpret transition probabilitiy matrices in Section 7.9.) However, the general pattern is consistent with the idea that those in the Sad and Disliked latent statuses are more likely to return to a Not Depressed latent status than those in either the Sad + Disliked or Depressed latent statuses.

7.8   EMPIRICAL EXAMPLE: DATING AND SEXUAL RISK BEHAVIOR

This example is drawn from a published article by Lanza and Collins (2008) that used LTA to examine dating and sexual risk behavior over time. The sample consisted of 2,937 participants in Rounds 2, 3, and 4 of the National Longitudinal Survey of Youth (NLSY), 1997 (Centers for Disease Control and Prevention, 2004) who were age 17 or 18 at Round 2 and did not report being married at Round 2, 3, or 4. Round 2 was conducted in 1998, when the participants were 17 or 18, Round 3 was conducted in 1999, and Round 4 was conducted in 2000. Four categorical variables were used as indicators of dating and sexual risk behavior. They are: Number of dating partners in the past year, with response categories of "0," "1," and "2 or more"; Had sex in past year, with response categories of "yes" and "no"; Number of sexual partners in past year, with response categories of "0," "1," and "2 or more"; and Exposed to a sexually transmitted infection (STI) in the past year (whether or not the participant reported at least one instance of intercourse without use of a condom in the past year, in other words, possible exposure to STIs), with response categories of "Yes" and "No." Information about missing data on the indicators and covariates may be found in Lanza and Collins (2008).

Table 7.11 summarizes the results of fitting latent transition models ranging from two to six latent statuses to the dating and sexual risk behavior data. Like Tables 7.6 and 7.9, Table 7.11 contains degrees of freedom and G^2 values, but not the p-values associated with the G^2's. The information criteria provide inconsistent messages about which model best balances parsimony and fit. According to the AIC, the model with six latent statuses is preferred, whereas according to the BIC, the model with five latent statuses is preferred. Ultimately, we chose the five-latent-status model shown in Table 7.12, based on conceptual appeal and parsimony.

Table 7.12 contains the item-response probabilities that formed the basis for the selection of labels for the five latent statuses. Those in the Nondaters latent status

Table 7.11   Summary of Information for Selecting Number of Latent Statuses of Dating and Sexual Risk Behavior (NLSY, Rounds 2–4, $N = 2,937$)

Number of Latent Statuses	Number of Parameters Estimated	G^{2*}	df	AIC	BIC	ℓ
2	17	5,403.4	46,638	5437.4	5539.2	−18,120.3
3	32	3,556.9	46,623	3620.9	3812.4	−17,197.1
4	51	3,171.1	46,604	3273.1	3578.3	−17,004.1
5	74	2,565.3	46,581	2713.3	3156.2	−16,701.2
6	101	2,360.4	46,554	2562.4	3166.9	−16,598.8

$^*p$-values not reported because the degrees of freedom are too large.

Table 7.12 Five-Latent-Status Model of Dating and Sexual Risk Behavior (NLSY, Rounds 2–4, $N = 2,937$)

	Latent Status				
	Non-Daters	Daters	Monogamous	Multi-partner Safe	Multi-partner Exposed
Latent status prevalences					
Time 1	.19	.29	.12	.23	.18
Time 2	.13	.23	.22	.21	.21
Time 3	.11	.18	.29	.17	.25
*Item-response probabilities**					
Number of dating partners in past year					
0	**.76**[†]	.01	.10	.05	.02
1	.18	.20	**.66**	.02	.05
2 or more	.06	**.79**	.24	**.93**	**.93**
Had sex in past year					
No	**.98**	**.99**	.00	.00	.00
Yes	.02	.01	**1.00**	**1.00**	**1.00**
Number of sexual partners in past year					
0	**1.00**	**1.00**	.00	.02	.00
1	.00	.00	**.97**	.34	.09
2 or more	.00	.00	.03	**.64**	**.91**
Exposed to STI in past year					
No	**1.00**	**1.00**	.40	**.82**	.19
Yes	.00	.00	**.60**	.18	**.81**

Probability of transitioning to... Conditional on... ...Time 1 latent status	...Time 2 latent status				
Nondaters	**.61**[‡]	.18	.08	.09	.03
Daters	.01	**.57**	.16	.21	.05
Monogamous	.05	.04	**.68**	.09	.14
Multipartner Safe	.04	.11	.21	**.54**	.11
Multipartner Exposed	.01	.03	.15	.00	**.81**

Probability of transitioning to... Conditional on... ...Time 2 latent status	...Time 3 latent status				
Nondaters	**.64**	.15	.15	.06	.00
Daters	.04	**.53**	.19	.16	.08
Monogamous	.04	.05	**.66**	.01	.24
Multipartner Safe	.04	.10	.14	**.57**	.15
Multipartner Exposed	.02	.01	.25	.00	**.72**

$\ell = -16,701.2$.

*Constrained equal across times.

[†] Item-response probabilities > .5 in bold to facilitate interpretation.

[‡] Diagonal transition probabilities in bold to facilitate interpretation.

were very likely to report zero dating partners in the past year, no sex in the past year, no sex partners in the past year, and no STI exposure in that year. Those in the Daters latent status had a high probability of reporting two or more dating partners, but were unlikely to report sex in the past year, any sex partners, or any exposure to STIs. Individuals in the Monogamous latent status were likely to report just one dating partner, and very likely to report having had sex in the past year with only one sex partner. However, 60 percent of those in this latent status reported having unprotected sex, and thus potentially were exposed to STIs. Those in the Multipartner Safe latent status were likely to report two or more dating and sexual partners in the past year, but had a high probability of having used a condom every time they had sex. Finally, individuals in the Multipartner Exposed latent status were very likely to report having had two or more dating and sexual partners in the past year, and had a very high probability of possible exposure to STIs with at least one sex partner.

The largest latent status at Time 1 was Daters, followed by Multipartner Safe. By Time 3 the Monogamous latent status was the most prevalent. The Multipartner Exposed latent status was the second smallest at Time 1, but by Time 3 was the second most prevalent.

Table 7.12 also contains the transition probability matrices for Times 1 and 2 and for Times 2 and 3. The diagonal elements of each transition probability matrix reflect the probability of membership in the same latent status at two consecutive times of measurement; for example, Nondaters at Time 1 had about a 61 percent chance of being in the Nondaters latent status again at Time 2, whereas individuals in the Multipartner Exposed latent status at Time 1 had an 81 percent chance of being in that latent status at Time 2. Off-diagonal elements represent the probability of transitioning from one latent status at the earlier time to a different latent status one year later. Nondaters at Time 1 who changed latent status were most likely to transition to the Daters latent status, and Daters who changed latent status tended to move to Monogamous and Multipartner Safe. Individuals in the Monogamous latent status had the highest probability of transitioning to the high-risk Multipartner Exposed latent status at a later time. This probability was .14 for Time 1 to Time 2, and .24 for Time 2 to Time 3. (In comparison, the probability of transitioning from the Multipartner Safe to the Multipartner Exposed latent status was .11 for Time 1 to Time 2 and .15 for Time 2 to Time 3.) This suggests that membership in the Monogamous latent status may be associated with future sexual risk behavior. Individuals in the high-risk Multipartner Exposed latent status who changed latent status membership over time were very likely to transition to the Monogamous latent status. Thus although they may not have transitioned to a latent status involving safe sex, they would have been exposed to STIs from just one partner rather than from multiple partners.

7.9 INTERPRETING WHAT A LATENT TRANSITION MODEL REVEALS ABOUT CHANGE

Transition probability matrices must be interpreted very carefully to determine what they do and do not reveal about change over time. An important factor in the interpretation of latent transition models, as in any statistical model of change, is the amount of time that elapses between occasions of measurement, particularly in relation to the exact interpretation of the latent statuses and the wording of any questionnaire or interview items that are used as indicators. We illustrate these points using the delinquency and depression examples.

Collins and Graham (2002) discussed the effect that the timing and spacing of observations in a longitudinal study (called *temporal design* by Collins, 2006) has on the conclusions that can be drawn from the data. They pointed out that when the time allowed to elapse between observations is long compared to the rapidity with which stage transitions occur, many transitions may occur between measurement occasions and consequently be missed.

For example, consider the transition probability matrix fit to the adolescent delinquency data and shown in Table 7.7. The transition probability matrix shows that about 3 percent of those who were Nondelinquents at Time 1 were in the Verbal Antagonists latent status at Time 2. However, it is inappropriate to conclude that these individuals transitioned directly into the Verbal Antagonists latent status. In the Add Health study the occasions of measurement were about one year apart, and participants were asked about delinquent behavior in the past year. This means that when an individual indicated, for example, lying to parents and publicly rowdiness in the past year, there is no information about whether or not these events occurred at about the same time. One could have occurred weeks or months before the other. Thus, based on these data it is impossible to know how many adolescents went directly to this latent status and how many went to the Liars latent status for a time and then transitioned from there into the Verbal Antagonists latent status. Because no latent status emerged that involved endorsing the item about public rowdiness without endorsing lying to parents, it seems likely that individuals are members of the Liars latent status for a time before they move to the Verbal Antagonists latent status. However, this cannot be confirmed.

The adolescent depression example was also drawn from the Add Health data set and therefore from the same occasions of measurement spaced one year apart. However, the items that serve as indicators for this example asked about depression symptoms in the last week. This makes the transition probability matrix difficult to interpret, because the answers to the questionnaire items could have changed many times between observations. For example, an individual who is in the Disliked latent status at Time 1 and in the Depressed latent status at Time 2 could have, in theory, moved around among any of the latent statuses many times between Times 1 and 2.

Similarly, there is an important difference in the interpretation of the diagonal elements of the transition probability matrix in the depression example compared to the delinquency example. In the delinquency example the diagonal elements of the transition probability matrix were interpreted as reflecting stability. For example, $\tau_{5_2|5_1}$ in Table 7.7 was interpreted as the probability that an individual in the General

Delinquents latent status at Time 1 would remain in that latent status until Time 2. Because in the delinquency example the indicators were questionnaire items asking whether individuals had engaged in the various activities in the past year, and the observations were spaced a year apart, the diagonal elements of the transition probability matrix could reasonably be interpreted as reflecting stability in latent status membership across two times.

On the other hand, because the indicators in the depression example asked about depressed feelings in the past week, and the occasions of measurement were spaced a year apart, the diagonal transition probabilities may not be interpreted as stability of latent status membership. After the Time 1 measurement an individual could have transitioned any number of times to other latent statuses and just happened to be in the Depressed latent status again at Time 2. Thus, in Table 7.10, $\tau_{5_2|5_1}$ is interpreted simply as the probability of membership in the Depressed latent status at Time 2, conditional on membership in this same latent status at Time 1.

More frequent measurement would have allowed better tracking of these transitions and possibly would have helped to make the transition probability matrix more readily interpretable. On the other hand, more frequent measurement would also have been more costly, may have been burdensome to participants, or may have had other detrimental effects on the research. Such trade-offs must be considered whenever a longitudinal study is planned. When analyzing longitudinal data, it is important to understand the speed at which individuals may change over time in the behavior or outcome of interest, and how the temporal design might limit the conclusions that can be made based on the statistical model. These issues are discussed in more detail in Collins and Graham (2002) and Collins (2006).

The primary point of this section is that when interpreting transition probabilities it is important to evaluate them in light of the spacing of occasions of measurement, the wording of any questionnaire or interview items that serve as indicators, and the exact specifications of the model that has been fit. It may be that unobserved transitions could have occurred at a high rate between occasions of measurement. If this is a possibility, there may be more movement between latent statuses than there appears to be based on the transition probability matrix. We recommend being particularly cautious about referring to the diagonal elements of the transition probability matrix in terms of stability, or interpreting them as the probability of remaining in a latent status. These issues about interpreting parameters that reflect change over time are relevant for all statistical models of longitudinal data.

7.10   PARAMETER RESTRICTIONS IN LTA

Parameter restrictions may be used in LTA in much the same way as they are used in LCA. Restrictions may be placed on the latent status prevalences, item-response probabilities, or transition probabilities in a latent transition model. In the latent transition models discussed in this book, any restrictions on latent status prevalences are imposed at Time 1 only because, as explained in Section 7.6.1, only Time 1 latent status prevalences are actually estimated; latent status prevalences for all subsequent times are computed based on other estimated parameters.

As discussed in Chapter 4, two different types of parameter restrictions are commonly used: parameters may be *fixed* or *constrained*. A parameter that is fixed to a particular value is not estimated. Before estimation begins, its value must be specified in the range 0 to 1. This fixed value is off limits to the estimation procedure. When parameters are constrained, they are placed in an equivalence set with other parameters. The estimation of all of the parameters in an equivalence set is constrained to be equal to the same value, which can be any value in the range 0 to 1. A single latent transition model may contain any combination of freely estimated, fixed, and constrained parameters.

Because a fixed parameter is not estimated, it does not contribute to the total number of parameters estimated, P. An equivalence set counts as a single estimated parameter irrespective of how many parameters comprise the set.

Below we discuss several applications of parameter restrictions that are useful in LTA. In Section 7.11 we demonstrate the use of restrictions on item-response probabilities, and in Section 7.12 we demonstrate the use of restrictions on transition probabilities.

7.11 TESTING THE HYPOTHESIS OF MEASUREMENT INVARIANCE ACROSS TIMES

In our view it is a good idea to constrain the item-response probabilities in LTA to be equal across times whenever it is reasonable to do so, that is, whenever measurement invariance across times can reasonably be assumed. There are conceptual and practical reasons for this.

The conceptual reason is that latent transition models are much easier to interpret if the item-response probabilities are identical across times. Recall that in Chapter 5 we discussed measurement invariance across groups and illustrated how to test the hypothesis of measurement invariance. If the item-response probabilities are identical across groups, then the interpretation of the latent classes is identical across groups. This means that any observed group differences in latent class prevalences can be interpreted simply as quantitative differences; certain latent classes are larger in some groups than in others. On the other hand, if the item-response probabilities are not identical across groups, then the meaning of the latent classes varies across groups. To the extent that this occurs, group comparisons of latent class prevalences become less straightforward. When interpreting differences in latent class prevalences, it becomes necessary to take into account any differences in the meaning of the latent classes at the same time. Of course, depending on the research questions at hand, sometimes these more qualitative differences can be interesting in and of themselves.

In much the same way, the transition probability matrix in LTA is easier to interpret if there is measurement equivalence across times. If the item-response probabilities are identical across times, the meaning of the latent statuses remains constant over time. This means that, for example, an element on the diagonal of the transition probability matrix reflects the probability of membership in latent status s at Time $t + 1$ conditional on membership in the same latent status s at Time t. To the extent that the item-response probabilities corresponding to latent status s change over time, the meaning of latent status s will change. Then it is no longer as clear how to interpret this transition probability, because along with interpreting quantitative change over time in latent status membership, it becomes necessary to interpret change over time in the meaning of the latent statuses. However, depending on the research questions at hand, this qualitative change may be interesting, particularly if it is developmentally meaningful.

The practical reason for restricting the item-response probabilities to be equal across times is to help stabilize estimation and improve identification. In latent transition models there can be a large number of item-response probabilities, particularly in models with more than two times. Imposing parameter restrictions over time can decrease the number of estimated parameters considerably.

7.11.1 Empirical example: Adolescent depression

The hypothesis that measurement invariance holds across times can be tested empirically by comparing two models. In one model, Model 1, the item-response probabilities are free to vary across times. The other model, Model 2, is identical to Model 1 except that the item-response probabilities are constrained to be equal across times. These models are statistically nested (see Chapter 4), so it is possible to conduct a G^2 difference test of the hypothesis of measurement invariance across times.

This hypothesis test is illustrated using the adolescent depression example. Recall that in the model that is presented in Table 7.10, the item-response probabilities were constrained to be equal across times. Table 7.13 shows how parameter restrictions can be set up to specify this model. These restrictions were used in Model 2. The results of comparing Models 1 and 2 are presented in Table 7.14. The G^2 difference test was not significant (difference $G^2 = 42.3$, $df = 40$, $p = .37$), indicating that the restrictions on the item-response probabilities did not result in a significant decrement in model fit. In addition, the AIC and BIC both pointed to the model in which the item-response probabilities were constrained equal across times. Taken together, these results suggest that we do not have to reject the hypothesis that measurement invariance holds across times.

Table 7.13   Parameter Restrictions Constraining Item-Response Probabilities to Be Equal Across Times for Adolescent Depression Example

	Latent Status				
	Not Depressed	Sad	Disliked	Sad + Disliked	Depressed
Time 1					
Response of Yes					
Couldn't shake blues	a	b	c	d	e
Felt depressed	f	g	h	i	j
Felt lonely	k	l	m	n	o
Felt sad	p	q	r	s	t
People unfriendly	u	v	w	x	y
Felt disliked	z	aa	bb	cc	dd
Life was failure	ee	ff	gg	hh	ii
Life not worth living	jj	kk	ll	mm	nn
Time 2					
Response of Yes					
Couldn't shake blues	a	b	c	d	e
Felt depressed	f	g	h	i	j
Felt lonely	k	l	m	n	o
Felt sad	p	q	r	s	t
People unfriendly	u	v	w	x	y
Felt disliked	z	aa	bb	cc	dd
Life was failure	ee	ff	gg	hh	ii
Life not worth living	jj	kk	ll	mm	nn

Note. Item-response probabilities designated with the same letter or pair of letters form an equivalence set.

Table 7.14   Fit Statistics for Test of Measurement Invariance Across Times for Adolescent Depression Example (Add Health Public-Use Data, Waves I and II; $N = 2,061$)

	G^2	*df*	AIC	BIC	ℓ
Model 1:					
Item-response probabilities vary across times	5,070.6	65,431	5,278.6	5,864.2	−14,513.5
Model 2:					
Item-response probabilities equal across times	5,112.9	65,471	5,240.9	5,601.3	−14,534.7
$G_2^2 - G_1^2 = 42.3, df = 40, p = .37$					

7.12   TESTING HYPOTHESES ABOUT CHANGE BETWEEN TIMES

Investigators may have a number of research questions that concern change over time. There may be a general question about whether or not there is a significant amount of change between Times 1 and 2. There may be a hypothesis that a transition between certain pairs of latent statuses is theoretically impossible. If the latent statuses can in some sense be considered to be ordered by severity, extremity, quality, or some other dimension, there may be a research question about whether change can be considered to be in a single direction, that is, all transitions reflecting an increase or all transitions reflecting a decrease.

Hypotheses like these that concern the nature or amount of change can be tested by placing restrictions on elements of the transition probability matrix, and then comparing the fit of the restricted model to that of a less restricted model. If the more restricted, and therefore simpler, model fits about as well as the less restricted model, then the simpler model is preferred. On the other hand, if the more restricted model provides a significantly worse fit to the data than the less restricted model, then the conclusion is that the additional parameters estimated in the less restricted model are necessary. This sort of hypothesis test directly informs scientific conclusions about change.

To illustrate how to express hypotheses about change in terms of restrictions on the transition probabilities, let us return to the adolescent depression example. Suppose that an investigator wishes to test the hypothesis that there is no change in depression between Times 1 and 2. This hypothesis can be tested by comparing two models. Model 1 is the model reported in Table 7.10. Model 2 is identical except that it contains a transition probability matrix that is restricted like the one in Table 7.15. In Table 7.15, the diagonal transition probability element is fixed to 1 for each latent status, and the remaining elements in each row are fixed to 0 (note that if an element in the row is fixed to 1 the others must all equal 0 by Equation 7.4). Thus in this model, no transition probabilities are estimated; all are fixed. This transition probability matrix expresses the idea that latent status membership at Time 2 is the same as at Time 1 for everyone.

The results of this hypothesis test are shown in Table 7.16. As the table shows, Model 2 had 20 more degrees of freedom than Model 1. This is because the entire transition probability matrix was fixed in Model 2, resulting in estimation of 20 fewer parameters. Table 7.16 shows that the fit of Model 2 to the data was much worse than

Table 7.15   Fixed Transition Probability Parameter Values Expressing a Model of No Change in the Five-Latent-Status Model of Adolescent Depression

	Latent Status				
	Not Depressed	Sad	Disliked	Sad + Disliked	Depressed
Probability of transitioning to... *Conditional on...*			*...Time 2 latent status*		
...Time 1 latent status					
Not Depressed	1	0	0	0	0
Sad	0	1	0	0	0
Disliked	0	0	1	0	0
Sad + Disliked	0	0	0	1	0
Depressed	0	0	0	0	1

Table 7.16   Fit Statistics for Test of Hypothesis That Latent Status Membership Is Identical at Times 1 and 2 for Adolescent Depression Example (Add Health Public-Use Data, Waves I and II; $N = 2{,}061$)

	G^2	df	AIC	BIC	ℓ
Model 1:					
Transition probability matrix freely estimated	5,112.9	65,471	5,240.9	5,601.3	−14,534.7
Model 2:					
Transition probability matrix fixed as in Table 7.15	6,188.3	65,491	6,276.3	6,524.1	−15,072.4
$G_2^2 - G_1^2 = 1{,}075.4, df = 20, p < .0001$					

that of Model 1. The AIC and BIC were larger in Model 2, and the difference G^2 was highly significant ($G_\Delta^2 = 1{,}075.4, df = 20, p < .0001$). Based on this hypothesis test, we reject the hypothesis that latent status membership at Time 2 is the same as latent status membership at Time 1.

Now suppose that an investigator wishes to test the hypothesis that the latent statuses Sad and Disliked represent distinctly different starting points for a process of depression, and that although it is possible to move from Sad to Sad + Disliked and from Disliked to Sad + Disliked, it is not possible to move from Sad to Disliked or from Disliked to Sad. The parameter restrictions in Table 7.17 are consistent with this hypothesis. All of the transition probabilities are freely estimated except $\tau_{\text{Disliked}_2|\text{Sad}_1}$ and $\tau_{\text{Sad}_2|\text{Disliked}_1}$, which are fixed to zero, indicating that these transitions are not possible. The hypothesis that $\tau_{\text{Disliked}_2|\text{Sad}_1} = \tau_{\text{Sad}_2|\text{Disliked}_1} = 0$ can be tested by comparing the fit of a model with parameter restrictions like those in Table 7.17 to the fit of the model shown in Table 7.10, in which all transition probabilities are freely estimated.

Table 7.17   Fixed Transition Probability Parameter Values Expressing a Model in Which There Is No Movement Between Sad and Disliked in the Five-Latent-Status Model of Adolescent Depression

	Latent Status				
	Not Depressed	Sad	Disliked	Sad + Disliked	Depressed
Probability of transitioning to...	...*Time 2 latent status*				
Conditional on...					
...*Time 1 latent status*					
Not Depressed	*	*	*	*	*
Sad	*	*	0	*	*
Disliked	*	0	*	*	*
Sad + Disliked	*	*	*	*	*
Depressed	*	*	*	*	*

Note. * denotes a free parameter.

Table 7.18 Fixed Transition Probability Parameter Values Expressing a Model in Which There Is Only Increasing Depression Across Time in the Five-Latent-Status Model of Adolescent Depression

	Latent Status				
	Not Depressed	Sad	Disliked	Sad + Disliked	Depressed
Probability of transitioning to...			*...Time 2 latent status*		
Conditional on...					
...Time 1 latent status					
Not Depressed	*	*	*	*	*
Sad	0	*	*	*	*
Disliked	0	0	*	*	*
Sad + Disliked	0	0	0	*	*
Depressed	0	0	0	0	1

Note.* denotes a free parameter.

As another example, suppose an investigator believes that the latent status Disliked reflects more advanced depression than the latent status Sad. Further suppose that the investigator wishes to specify that it is impossible to transition from a more depressed latent status at Time 1 into a less depressed latent status at Time 2. The parameter restrictions shown in Table 7.18 are consistent with this hypothesis. The hypothesis can be tested using essentially the same approach as that described above. The fit of a model with parameter restrictions like those in Table 7.18 would be compared to the fit of the model reported in Table 7.10, in which all transition probabilities are freely estimated. In this case the difference in degrees of freedom would equal 10; we leave it as an exercise for the reader to determine why.

7.13 RELATION BETWEEN RMLCA AND LTA

In this chapter we have discussed both RMLCA and LTA approaches to analysis of longitudinal data. In this section we discuss how these two models are related. Throughout the discussion we assume that any restrictions on item-response probabilities are implemented in the same way in the RMLCA and LTA models. We also assume that no restrictions have been placed on the latent class prevalences in the RMLCA or on the latent status prevalences or transition probabilities in the LTA.

7.13.1 Relation between RMLCA and LTA when there are two times

In general, a latent transition model that involves two times and S latent statuses at each time, and that has no restrictions on the latent status prevalences or transition probabilities, can be fit equivalently as a repeated measures latent class model with S^2 latent classes. Each latent class prevalence in the RMLCA corresponds to the unconditional probability of membership in latent status s_1 at Time 1 and latent status s_2 at Time 2. For example, if the adolescent depression model were fit as a repeated measures latent class model with $5^2 = 25$ latent classes, one of the latent classes would correspond to the unconditional probability of membership in the Sad latent status at Time 1 and the Sad + Disliked latent status at Time 2.

The latent class prevalences in a RMLCA with two times can be obtained from the latent status prevalences and the transition probabilities in a corresponding LTA. Let $\gamma_{s_1 s_2}$ represent the latent class prevalence corresponding to the probability of membership in latent status s_1 at Time 1 and latent status s_2 at Time 2. Then

$$\gamma_{s_1 s_2} = \delta_{s_1} \tau_{s_2 | s_1} \tag{7.12}$$

and it follows that

$$\tau_{s_2 | s_1} = \frac{\gamma_{s_1 s_2}}{\delta_{s_1}}. \tag{7.13}$$

The number of latent class prevalences estimated in the RMLCA with S^2 latent classes is identical to the total number of latent status prevalences and transition probabilities estimated in the equivalent LTA with S latent statuses. In RMLCA $S^2 - 1$ latent class prevalences are estimated (because by Equation 2.1 the vector of latent class prevalences sums to 1). In LTA, $S - 1$ latent status prevalences are estimated at Time 1 (Equations 7.1 and 7.9), and (when $T = 2$) $S(S - 1)$ transition probabilities are estimated (Equations 7.4 and 7.11), for a total of $S(S - 1) + S - 1 = S^2 - 1$ estimated parameters.

Thus when $T = 2$ and the assumptions about parameter restrictions stated at the beginning of this section are met, LTA is a reparameterization of RMLCA. In other words, under these conditions the models are mathematically equivalent, and the parameters in one are simple functions of the parameters in the other. There are two reasons for going to the trouble of creating the reparameterization used in LTA. First, depending on the research questions motivating the analysis, it can be more conceptually appealing and helpful to interpret the LTA parameters than the RMLCA parameters. Second, LTA facilitates placement of parameter restrictions directly on the transition probabilities. As was seen above, this can be essential in testing certain hypotheses about development.

7.13.2   Relation between RMLCA and LTA when there are three or more times

In this section we explore the correspondence between RMLCA and LTA when $T \geq 3$. Here the correspondence between typical RMLCA and LTA is less straightforward. As we did in the preceding section, we assume that any restrictions on item-response probabilities are implemented in the same way in the RMLCA and LTA models, and that no restrictions have been placed on the latent class prevalences in the RMLCA or on the latent status prevalences or transition probabilities in LTA.

In LTA there are $T - 1$ transition probability matrices, so in addition to the $S - 1$ latent status prevalences there are $(T - 1)S(S - 1)$ transition probabilities estimated (Equation 7.11). For example, suppose that instead of two times in the adolescent depression example there were $T = 3$ times. Then a standard latent transition model fit to the depression data would include two transition probability matrices, one expressing latent status membership at Time 2 conditional on Time 1, and another matrix expressing latent status membership at Time 3 conditional on Time 2. There would be $2 \times 5 \times 4 = 40$ transition probabilities estimated, plus four latent status prevalences at Time 1, for a total of 44 latent status prevalence and transition probability parameters estimated.

However, still supposing that the depression data included three times, a RMLCA could include up to S^T latent classes, that is, a latent class corresponding to each possible combination of Time 1, Time 2, and Time 3 latent status memberships. In the adolescent depression example there would be $5^3 = 125$ latent classes, so with no parameter restrictions on the latent class prevalences, 124 latent class prevalences would be estimated. Thus the RMLCA would estimate up to 80 additional parameters.

Why does LTA with S latent statuses estimate fewer parameters than the corresponding unrestricted RMLCA with S^2 latent classes when $T > 2$? The reason is that when $T > 2$, the repeated measures latent class model and the latent transition model are not directly equivalent. Because LTA is a type of first-order Markov model (Everitt, 2006), latent transition models make the assumption that all the information available to predict Time $t + 1$ is available at Time t. In other words, it is sufficient to condition transition probabilities on the immediately earlier time only and not to condition on any additional times. This assumption does not have any impact on the analysis when $T = 2$, because in this case Time t is the only time available for predicting Time $t + 1$. However, when $T > 2$, the assumption becomes important.

If the Markov assumption is not made, it is in theory possible to condition the transition probability parameters on any available earlier times, which would make LTA and RMLCA equivalent. If this were done in the hypothetical example in which the adolescent depression study involves three times, the Time 1 to Time 2 transition probability matrix would be the same, but the second transition probability matrix would be made up of the probabilities of latent status membership at Time 3 conditional on latent status membership at BOTH Times 1 and 2. This new transition probability matrix would be larger than the Time 1 to Time 2 transition probability matrix. It would have S columns (corresponding to Time 3 latent status membership), but now would have S^2 rows (corresponding to latent status membership at both Times 1 and 2). Each element could be represented by $\tau_{s_3|s_1,s_2}$. The number of transition probabilities estimated would now be $S(S - 1) = 20$ in the first transition probability matrix and $S^2(S - 1) = 100$ in the second. When added to the four latent status prevalences that would be estimated, this comes to 124 parameters, which is the same as the number of latent class prevalences estimated in the corresponding RMLCA described in the preceding paragraph.

There are fewer parameters estimated in the standard LTA than in the RMLCA because the Markov assumption in effect places restrictions on certain parameters. The Markov assumption, put in LTA terms, is

$$\tau_{s_3|s_1,s_2} = \tau_{s_3|s_2} \tag{7.14}$$

for all $s_1 = 1, ..., S$. Thus the $\tau_{s_3|s_1,s_2}$ parameters are in a sense present in the LTA model, but they are placed in equivalence sets corresponding to Time 1 latent status membership so that only $\tau_{s_3|s_2}$ parameters are estimated.

Latent transition models in which transitions are conditioned on more than one time are outside the scope of this book, but it is possible to fit them in some cases. Care must be taken in model selection and interpretation, as identification problems are common.

7.13.3   When to use RMLCA versus LTA

Except for the situation discussed above in which the two models are equivalent, RMLCA and LTA are subtly different models. RMLCA is suitable for research questions involving subgroups characterized by their patterns of change across several time points. It usually does not enable examination of all possible transitions between latent classes between adjacent pairs of times. In contrast, LTA is suitable for questions involving transitions between latent statuses across adjacent time points. It does not enable modeling of subgroups corresponding to the change across three or more times undergone by individuals.

Above we showed that a latent transition model with S latent statuses estimates fewer parameters than a latent class model with S^T latent statuses when $T \geq 3$. However, it is not necessarily true that, in general, LTA estimates fewer parameters than RMLCA. The reason is that for most RMLCA's, all S^T latent classes will not be needed. Instead, usually a smaller number of latent classes, each corresponding to a distinct pattern of change over time, represents the data well. If the number of latent classes is less than $S - 1 + (T - 1)S(S - 1)$, then, all else being equal, the number of parameters estimated in the RMLCA will be less than the number estimated in the LTA.

This means that if change is characterized by a few distinct patterns, often RMLCA can estimate considerably fewer parameters than can LTA. On the other hand, if transitions can occur between any pair of latent statuses and the first-order Markov assumption (the assumption that Time $t + 1$ latent status membership is conditional on latent status membership at Time t only, and not on latent status membership at any earlier time) is reasonable, LTA can estimate considerably fewer parameters than RMLCA.

7.14 INVARIANCE OF THE TRANSITION PROBABILITY MATRIX WHEN THERE ARE THREE OR MORE TIMES

When $T \geq 3$ it may be of interest to test the hypothesis that the transition probabilities are identical across times; that is $\tau_{s_2|s_1} = \tau_{s_3|s_2}\ldots = \ldots\tau_{s_T|s_{T-1}}$. In most applications this hypothesis makes sense only when the amount of time allowed to elapse between Time t and Time $t + 1$ is at least approximately the same for all $t, t + 1$.

Consider the dating and sexual risk behavior example shown in Table 7.12, which involves three times. There is some evidence that the transition probabilities differ across times. For example, the probability of a Nondater transitioning to the Monogamous latent status between Times 1 and 2 was .08, whereas this transition had a probability of .15 between Times 2 and 3. An individual in the high-risk Multipartner Exposed latent status had a .15 probability of transitioning to the Monogamous latent status between Times 1 and 2, and a .25 probability between Times 2 and 3. These differences and others suggest that it may be interesting to test the null hypothesis that the two transition probability matrices are equal.

The null hypothesis of equal transition probabilities across the three times can be tested by comparing the fit of two models. In Model 1 the Time 1 to Time 2 transition probabilities are free to be different from the Time 2 to Time 3 transition probabilities. This is the model that is shown in Table 7.12. In Model 2 the Time 1 to Time 2 transition probabilities are constrained to be equal to the Time 2 to Time 3 transition probabilities. Table 7.19 shows how the parameter restrictions on the transition probability matrices could be specified in Model 2. Table 7.20 shows the estimates of transition probabilities obtained from fitting Model 2 to the dating and sexual risk behavior example.

A G^2 difference test comparing Model 1 to Model 2 can be conducted in much the same way as other G^2 difference tests have been described in this book. The test has degrees of freedom equal to the difference in the number of parameters estimated in the two models. The fit statistics for Models 1 and 2 and the difference test are shown in Table 7.21. The hypothesis test was significant ($G^2_\Delta = 36.9, df = 20, p = .01$), and the AIC and BIC both favored the model that allows the two transition probability matrices to vary. Therefore, we reject the null hypothesis and conclude that the transition probabilities from Time 1 to Time 2 and the transition probabilities from Time 2 to Time 3 are different.

7.15 SUGGESTED SUPPLEMENTAL READINGS

Lanza and Collins (2006) contains a more detailed account of the RMLCA of heavy drinking behavior summarized in this chapter. Another application of RMLCA can

Table 7.19   Parameter Restrictions Constraining Transition Probabilities to Be Equal Across Three Times in a Hypothetical Five-Latent-Status Model of Adolescent Depression

	Latent Status				
	Non-daters	Daters	Monogamous	Multi-partner Safe	Multi-partner Exposed
Probability of transitioning to... *Conditional on...* *...Time 1 latent status*			*...Time 2 latent status*		
Nondaters	a	b	c	d	e
Daters	f	g	h	i	j
Monogamous	k	l	m	n	o
Multipartner Safe	p	q	r	s	t
Multipartner Exposed	u	v	w	x	y
Probability of transitioning to... *Conditional on...* *...Time 2 latent status*			*...Time 3 latent status*		
Nondaters	a	b	c	d	e
Daters	f	g	h	i	j
Monogamous	k	l	m	n	o
Multipartner Safe	p	q	r	s	t
Multipartner Exposed	u	v	w	x	y

Note. Transition probabilities designated with the same letter form an equivalence set.

Table 7.20   Five-Latent-Status Model of Dating and Sexual Risk Behavior with Transition Probability Matrices Constrained Equal Across Three Times (NLSY, Rounds 2–4, $N = 2{,}937$)

	Latent Status				
	Non-daters	Daters	Monogamous	Multi-partner Safe	Multi-partner Exposed
Probability of transitioning to... *Conditional on...* *...Time t latent status*			*...Time $t + 1$ latent status*		
Nondaters	**.62**[*]	.18	.11	.08	.02
Daters	.02	**.55**	.17	.19	.06
Monogamous	.04	.05	**.67**	.05	.19
Multipartner Safe	.04	.10	.17	**.56**	.13
Multipartner Exposed	.01	.02	.21	.00	**.76**

$\ell = -16{,}719.7$.

[*]Diagonal transition probabilities in bold to facilitate interpretation.

be found in Jackson, Sher, and Wood (2000). These authors used RMLCA to identify longitudinal types of alcohol and tobacco use disorders.

Lanza, Flaherty, and Collins (2003) and Collins and Flaherty (2002) provide readable introductions to LTA. Lanza and Collins (2008) includes the example on dating

Table 7.21 Fit Statistics for Test of Invariance of Transition Probability Matrices in the Dating and Sexual Risk Behavior Example (NLSY, Rounds 2–4, $N = 2,937$)

	G^2	df	AIC	BIC	ℓ
Model 1:					
Transition probability matrices vary across times	2,565.3	46,581	2,713.3	3,156.2	−16,701.2
Model 2:					
Transition probability matrices equal across times	2,602.2	46,601	2,710.2	3,033.4	−16,719.7
$G_2^2 - G_1^2 = 36.9, df = 20, p = .01$					

and sexual risk behavior discussed in the current chapter, and Lanza et al. (2003) includes the depression subtypes example.

Here we list a few additional examples of applications of LTA in which readers may be interested. Auerbach and Collins (2006) applied LTA to data on alcohol use during the transition to adulthood. Collins (2002) used LTA to examine the gateway hypothesis of drug use onset. Langeheine, Stern, and van de Pol (1994) modeled gaining the ability to solve math problems. Posner, Collins, Longshore, and Anglin (1996) examined safer sexual behaviors among injection drug users. Patrick et al. (2009) modeled substance use onset in a sample of South African adolescents. Reboussin, Liang, and Reboussin (1999) modeled transitions in health status among older adults.

7.16 POINTS TO REMEMBER

• RMLCA is an approach to analysis of longitudinal data in which the latent classes correspond to different patterns of categorical or discrete change over time.

• Because RMLCA does not fit a functional form to change over time, it is particularly useful when change over time is not well represented by any particular functional form.

• LTA can be used to model change between two adjacent times. Latent transition models estimate three different sets of parameters: item-response probabilities at each time; latent status prevalences; and latent status transitions.

• LTA models are usually fit to very large contingency tables, because repeated observations over time on several variables are involved. Because the contingency tables are large, they tend to be sparse. This sparseness renders the G^2 statistic of limited utility in model selection. Investigators usually rely on penalized fit statistics such as the AIC and BIC. In addition, examining the latent class structure within each time point can be informative in the selection of an LTA model.

• Parameter restrictions may be used in LTA in much the same way as they are used in LCA. Restrictions may be placed on any of the three sets of parameters in LTA. Parameter restrictions on the transition probabilities are often particularly useful in testing models of change over time.

• It is a good idea to constrain the item-response probabilities to be equal across times in LTA whenever it is reasonable to do so. This helps to avoid identification problems and makes the results more interpretable in most situations.

• Investigators should be cautious when interpreting the transition probabilities in LTA. Interpretation should include reference to the amount of time that has elapsed between observations and how time is referenced in the indicators of the latent variable.

7.17 WHAT'S NEXT

In the next chapter, where we continue our treatment of LTA, the latent transition model is extended in two directions. First, we discuss multiple-group LTA; then, we discuss LTA with covariates.

CHAPTER 8

MULTIPLE-GROUPS LATENT TRANSITION ANALYSIS AND LATENT TRANSITION ANALYSIS WITH COVARIATES

8.1 OVERVIEW

In Chapter 5 we discussed how to incorporate a grouping variable into LCA, in order to examine differences between observed groups. In Chapter 6 we discussed how to incorporate covariates into LCA, in order to predict latent class membership. These ideas can be extended readily to LTA (Humphreys and Janson, 2000; Reboussin, Liang, and Reboussin, 1999; Reboussin, Reboussin, Liang, and Anthony, 1998; Vermunt, Langeheine, and Böckenholt, 1999). Grouping variables can be used to examine group differences in item-response probabilities, latent status prevalences, and/or transition probabilities. In the models discussed in this chapter, covariates can be used to predict latent status membership and/or latent status transitions. Grouping variables and covariates may be incorporated together in the same model, in which case group differences in the prediction of latent status membership and/or latent status transitions can be modeled. In this chapter we discuss how to include grouping variables and covariates in latent transition models.

8.2   LTA WITH A GROUPING VARIABLE

Just as in LCA, LTA sometimes involves examining differences between populations, where population membership can be measured directly and the available sample can accordingly be divided into population groups. For example, there may be interest in gender differences, age differences, cohort differences, or differences between experimental conditions.

Group similarities and differences may be manifested in any of the three sets f parameters in latent transition models. When a grouping variable is included in a latent transition model, several types of group differences may be examined. It is possible to test for measurement invariance across groups by testing the null hypothesis that the item-response probabilities are identical across groups. It is also possible to test the equivalence of the latent status prevalences at Time 1, and the equivalence of the transition probabilities, across groups.

Many of the basic considerations relevant to multiple-group analyses are the same in LTA as they are in LCA, so we refer the reader to Chapter 6 for an in-depth treatment of model selection when there are multiple groups and measurement invariance across groups. In this chapter we concentrate on aspects of multiple-group analyses that are particular to LTA.

8.2.1   Empirical example:
Adolescent depression

Let us return to the empirical example of LTA on adolescent depression data that was introduced in Chapter 7. In this example there are five latent statuses of depression: Not Depressed, Sad, Disliked, Sad + Disliked, and Depressed. Recall that the data set involves two cohorts, Grades 10 and 11, assessed at two time points. To what extent are the latent status prevalences and transition probabilities the same or different across these two cohorts? This question can be addressed using multiple-group LTA.

Table 8.1 shows the results of LTA that used cohort as a grouping variable. The table shows the latent status prevalences and transition probabilities for each cohort. Overall, the parameter estimates look roughly similar across the groups. In a later section we demonstrate how to test the null hypothesis that the Grades 10 and 11 cohorts have identical latent status prevalences and transition probabilities. First, let us discuss the LTA model and notation when a grouping variable is included.

8.3   MULTIPLE-GROUP LTA:
MODEL AND NOTATION

As we did in Chapter 7, suppose that there are $j = 1, ..., J$ observed indicator variables measuring the latent statuses, and the J observed variables have been

Table 8.1 Latent Status Prevalences and Transition Probabilities for Five-Latent-Status Model of Adolescent Depression, by Cohort (Add Health Public-Use Data, Waves I and II; $N = 2,061$)

	Latent Status				
	Not Depressed	Sad	Disliked	Sad + Disliked	Depressed
Latent status prevalences					
Grade 10 cohort					
Time 1	.37	.20	.20	.11	.13
Time 2	.39	.25	.15	.12	.10
Grade 11 cohort					
Time 1	.35	.26	.17	.12	.09
Time 2	.41	.22	.16	.11	.10
Grade 10 cohort					
Probability of transitioning to...		*...Time 2 latent status*			
Conditional on...					
...Time 1 latent status					
Not Depressed	**.75**[†]	.12	.09	.02	.01
Sad	.21	**.67**	.04	.04	.05
Disliked	.26	.05	**.45**	.18	.05
Sad + Disliked	.02	.41	.13	**.38**	.06
Depressed	.10	.13	.03	.19	**.54**
Grade 11 cohort					
Probability of transitioning to...		*...Time 2 latent status*			
Conditional on...					
...Time 1 latent status					
Not Depressed	**.76**	.12	.09	.02	.01
Sad	.37	**.45**	.05	.09	.04
Disliked	.23	.05	**.51**	.16	.06
Sad + Disliked	.06	.26	.18	**.37**	.14
Depressed	.02	.20	.07	.07	**.64**

$\ell = -14,521.2$.

[†] Diagonal transition probabilities in bold to facilitate interpretation.

Note. Item-response probabilities constrained equal across times.

measured at $t = 1, ..., T$ times. In the adolescent depression example, $J = 8$ and $T = 2$. Observed variable j has R_j response categories; in other words, to keep the exposition simple we assume that the number of response categories for observed variable j does not vary across times. In the example, $R_j = 2$ for all observed variables. At time t, $r_{j,t} = 1, ..., R_j$ indexes the individual response categories. In addition, we include a grouping variable V with $q = 1, ..., Q$ groups. In the example V represents cohort, and there are $Q = 2$ groups: Grades 10 and 11.

The contingency table formed by cross-tabulating the J indicator variables measured at T times and the grouping variable V has $W = Q \prod_{t=1}^{T} \prod_{j=1}^{J} R_j$ cells. In the adolescent depression example, $W = 2 \times 2^{2 \times 8} = 2 \times 2^{16} = 131,072$. Corresponding to each of the W cells in the contingency table is a complete response pattern, which is a vector of responses to the grouping variable and the J indicator variables at T times, represented by $\mathbf{y} = (q, r_{1,1}, ..., r_{J,T})$. Let $\mathbf{Y}$ refer to the array of response patterns. Each response pattern $\mathbf{y}$ in the array is associated with probability of occurrence $P(\mathbf{Y} = \mathbf{y})$, and within each group q, $\sum P(\mathbf{Y} = \mathbf{y}|V = q) = 1$.

Let L represent the categorical latent variable with S latent statuses, where $s_1 = 1, ..., S$ at Time 1, $s_2 = 1, ..., S$ at Time 2, and so on, up to $s_T = 1, ..., S$ at Time T. (To simplify the exposition we assume that the number of latent statuses is identical across groups.) In the example the latent variable L is adolescent depression, and there are $S = 5$ latent statuses at each time. Finally, $I(y_{j,t} = r_{j,t})$ is an indictor function that equals 1 when the response to variable j at Time $t = r_{j,t}$ and equals 0 otherwise. (As mentioned previously, this function is merely a device for picking out the appropriate ρ parameters to multiply together.) Then

$$P(\mathbf{Y} = \mathbf{y}|V = q) = \sum_{s_1=1}^{S} \cdots \sum_{s_T=1}^{S} \delta_{s_1|q} \tau_{s_2|s_1,q} \cdots \tau_{s_T|s_{T-1},q} \prod_{t=1}^{T} \prod_{j=1}^{J} \prod_{r_{j,t}=1}^{R_j} \rho_{j,r_{j,t}|s_t,q}^{I(y_{j,t}=r_{j,t})},$$

(8.1)

where $\delta_{s_1|q}$ is the probability of membership in latent status s_1 at Time 1, conditional on membership in group q; $\tau_{s_2|s_1,q}$ is the probability of membership in latent status s_2 at Time 2 conditional on membership in latent status s_1 at Time 1 and membership in group q; and $\rho_{j,r_{j,t}|s_t,q}$ is the probability of response $r_{j,t}$ to item j at Time t, conditional on membership in latent status s_t at Time t and membership in group q.

In other words, all of the parameters are now conditioned on group membership. This means that the model includes separate estimates of the latent status prevalences, transition probabilities, and item-response probabilities for each group. This makes it possible to compare these estimates across groups. In the example, this means that it is possible to compare latent status prevalences, transition probabilities, and item-response probabilities across cohorts.

8.4   COMPUTING THE NUMBER OF PARAMETERS ESTIMATED IN MULTIPLE-GROUP LATENT TRANSITION MODELS

As Equation 8.1 shows, in the multiple-group latent transition model the latent status prevalences, transition probabilities, and item-response probabilities are conditioned on group membership, in essentially the same way as the latent class prevalences and item-response probabilities are conditioned on group membership in multiple-group LCA.

This means that if no parameter restrictions are specified, the number of δ, τ, and ρ parameters estimated is multiplied by the number of groups Q. Thus there are $Q(S-1)$ δ's, $Q(T-1)S(S-1)$ τ's, and $QST \sum_{j=1}^{J}(R_j - 1)$ ρ's estimated. If parameter restrictions are specified, the number of parameters estimated is reduced.

8.5 HYPOTHESIS TESTS CONCERNING GROUP DIFFERENCES IN LATENT STATUS PREVALENCES AND TRANSITION PROBABILITIES: GENERAL CONSIDERATIONS

Using multiple-group LTA it is possible to test the hypothesis that the latent status prevalences at Time 1 are equivalent across groups and the hypothesis that the transition probabilities are equivalent across groups. It is also possible to conduct an overall test that both of these sets of parameters are equivalent across groups. The same general strategy of identifying nested models and comparing their fit using the difference G^2 test that has been described elsewhere in this volume is used to perform these hypothesis tests (see Chapters 4 and 5 for more detail). Tests of group equivalence of latent status prevalences and/or transition probabilities make the most sense when measurement equivalence across both groups and times either has been established or can confidently be assumed, and accordingly the item-response probabilities are constrained to be equal across groups and times.

Table 8.2 illustrates four different multiple-group latent transition models representing different types and degrees of equivalence across groups. In all four models the item-response probabilities are equal across groups and times. In Model 1, both latent status prevalences at Time 1 and latent status transitions are free to vary across groups. In Model 2, both of these sets of parameters are constrained to be equal across groups. Model 3 specifies that the latent status prevalences at Time 1 are free to vary across groups, but the latent status transitions are constrained to be equal across groups. Model 4 specifies that the latent status transitions are free to vary across groups but the latent status prevalences at Time 1 are constrained to be equal across groups. In Sections 8.6 through 8.8 we demonstrate the use of these four basic types of models to test a variety of hypotheses concerning equivalence across groups.

Table 8.2   Varying Types of Equivalence Across Groups in Latent Transition Models

	Item-Response Probabilities	Latent Status Prevalences at Time 1	Transition Probabilities
Model 1	equal across groups and times	free	free
Model 2	equal across groups and times	equal across groups	equal across groups
Model 3	equal across groups and times	free	equal across groups
Model 4	equal across groups and times	equal across groups	free

8.6 OVERALL HYPOTHESIS TESTS ABOUT GROUP DIFFERENCES IN LTA

Suppose that an overall test of the equivalence of the latent status prevalences and transition probabilities across groups is desired. This can be accomplished by comparing Models 1 and 2. If the latent status prevalences and transition probabilities are equal across groups, Model 2 will fit the data about as well as Model 1. Thus the null hypothesis H_0 can be expressed as follows: "Model 2 fits as well as Model 1" or "The latent status prevalences and transition probabilities are not different across groups." This hypothesis implies that the parameter restrictions imposed in Model 2 are plausible.

A test of H_0 is provided by the likelihood-ratio test $G_{\Delta}^2 = G_2^2 - G_1^2$, which in theory is distributed as a χ^2 with $df = df_2 - df_1$. Just as in LCA, in LTA it is frequently helpful to examine the AIC and BIC in addition to the likelihood ratio statistic G_{Δ}^2 when deciding whether or not to conclude that the more restrictive model fits the data sufficiently well. This may be particularly useful when a large number of parameters are involved, as in practice the distribution of the G^2 difference may not be approximated well by the χ^2 when there are many df. As usual, lower AIC and BIC values are associated with the preferred model. A significant hypothesis test, that is, a rejection of H_0, suggests that at least one δ or τ parameter is different across groups. If H_0 is not rejected, the conclusion is that the δ and τ parameters can be treated as identical across groups in any subsequent latent transition analyses.

8.6.1 Empirical example: Cohort differences in adolescent depression

Let us follow the procedure outlined above to test the hypothesis that both the latent status prevalences and transition probabilities are equal across cohorts in the adolescent depression example. This requires fitting Models 1 and 2 from Table 8.2 to the adolescent depression data. Both models involve cohort as a grouping variable, and item-response probabilities are constrained to be equal across groups and times. Model 1, in which the latent status prevalences and transition probabilities are free to vary across cohorts, was shown in Table 8.1. Model 2 specifies that all of the parameters in the latent transition model are identical across cohorts.

Table 8.3 shows one way to specify equivalence sets for Model 2 for the latent status prevalences and transition probabilities. (Information about how to specify equivalence sets for item-response probabilities may be found in Chapters 5 and 7.) Note that the latent status prevalences are constrained to be equal across cohorts only at Time 1. As discussed in Chapter 7, in the latent transition models covered in this book, latent status prevalences for Times 2 through T are not estimated; rather,

Table 8.3   Parameter Restrictions Constraining Latent Status Prevalences and Transition Probabilities to Be Equal Across Cohorts in Adolescent Depression Example

	Latent Status				
	Not Depressed	Sad	Disliked	Sad + Disliked	Depressed
Latent status prevalences					
Grade 10 cohort					
Time 1	a	b	c	d	e
Grade 11 cohort					
Time 1	a	b	c	d	e
Grade 10 cohort					
Probability of transitioning to...		...*Time 2 latent status*			
Conditional on...					
...*Time 1 latent status*					
Not Depressed	f	g	h	i	j
Sad	k	l	m	n	o
Disliked	p	q	r	s	t
Sad + Disliked	u	v	w	x	y
Depressed	z	aa	bb	cc	dd
Grade 11 cohort					
Probability of transitioning to...		...*Time 2 latent status*			
Conditional on...					
...*Time 1 latent status*					
Not Depressed	f	g	h	i	j
Sad	k	l	m	n	o
Disliked	p	q	r	s	t
Sad + Disliked	u	v	w	x	y
Depressed	z	aa	bb	cc	dd

Note. Latent status prevalences and item-response probabilities designated with the same letter or pair of letters form an equivalence set.

they are computed from other estimated parameters. For this reason, they cannot be restricted. However, if the Time 1 latent status prevalences and the Time 1 to Time 2 transition probabilities are constrained to be equal across groups, it follows that the Time 2 latent status prevalences will be equal across groups (Equation 7.7).

Table 8.3 is only one example of how parameter restrictions could be specified. If desired, there could be additional restrictions on the latent status prevalences or transition probabilities. For example, some of the transition probabilities could be fixed at zero. If additional parameter restrictions are specified, they must be identical across the groups being compared in order to assess group differences in the latent status prevalences at Time 1 and transition probabilities from Time 1 to Time 2.

The parameter estimates and ℓ obtained from fitting Model 2 are shown in Table 8.4. Table 8.5 shows the Model 1 and 2 G^2's, degrees of freedom, AICs, BICs, and ℓ's, as well as the G^2 difference test of the hypothesis of no cohort differences in the latent status prevalences and transition probabilities. The G^2 difference test was not significant ($G^2_\Delta = 26.9, df = 24, p = .31$), suggesting that the latent status prevalences and transition probabilities were not significantly different across groups.

Table 8.4 Latent Status Prevalences and Transition Probabilities Constrained Equal Across Cohorts for Five-Latent-Status Model of Adolescent Depression (Add Health Public-Use Data, Waves I and II; $N = 2,061$)

	Latent Status				
	Not Depressed	Sad	Disliked	Sad + Disliked	Depressed
Latent status prevalences					
Grade 10 cohort					
Time 1	.36	.23	.18	.11	.11
Time 2	.40	.23	.15	.11	.10
Grade 11 cohort					
Time 1	.36	.23	.18	.11	.11
Time 2	.40	.23	.15	.11	.10
Grade 10 cohort					
Probability of transitioning to...		...*Time 2 latent status*			
Conditional on...					
...*Time 1 latent status*					
Not Depressed	**.75**[†]	.12	.09	.02	.02
Sad	.30	**.54**	.04	.07	.04
Disliked	.25	.05	**.47**	.17	.06
Sad + Disliked	.03	.34	.17	**.36**	.11
Depressed	.08	.16	.03	.14	**.59**
Grade 11 cohort					
Probability of transitioning to...		...*Time 2 latent status*			
Conditional on...					
...*Time 1 latent status*					
Not Depressed	**.75**	.12	.09	.02	.02
Sad	.30	**.54**	.04	.07	.04
Disliked	.25	.05	**.47**	.17	.06
Sad + Disliked	.03	.34	.17	**.36**	.11
Depressed	.08	.16	.03	.14	**.59**

$\ell = -14,534.7$.

[†] Diagonal transition probabilities in bold to facilitate interpretation.

Note. Item-response probabilities constrained equal across times.

In addition, the AIC and BIC both suggest that Model 2 provided a more optimal balance of model fit and parsimony. Therefore, we can comfortably conclude that there is no evidence of differences in the latent status prevalences and item-response probabilities across Grades 10 and 11.

If the null hypothesis had been rejected, the conclusion would have been that one or more of the latent status prevalences or transition probabilities differed across cohorts. It is possible to test more specific hypotheses about which set of parameters differs across groups. This is demonstrated in Section 8.6.2.

Table 8.5   Fit Statistics for Test of Cohort Differences in Latent Status Prevalences
and Transition Probabilities for Adolescent Depression Example (Add Health
Public-Use Data, Waves I and II; $N = 2,061$)

	G^2	df	AIC	BIC	ℓ
Model 1:					
Latent status prevalences and transition probabilities					
vary across cohorts	6,519.9	130,983	6,695.9	7,191.4	−14,521.2
Model 2:					
All parameters					
equal across cohorts	6,546.8	131,007	6,674.8	7,035.2	−14,534.7
$G_2^2 - G_1^2 = 26.9, df = 24, p = .31$					

Note. Models 1 and 2 are described in Table 8.2.

8.6.2 Empirical example: Gender differences in adolescent depression

As another example, let us test for gender differences in latent status prevalences
and transition probabilities in the adolescent depression example. (Here the Grades
10 and 11 cohorts are combined.) The overall null hypothesis that the latent status
prevalences and transition probabilities are equal across genders can be performed
in the same manner as the hypothesis test about equivalence across cohorts was
performed above, that is, by comparing Models 1 and 2 from Table 8.2.

Model 1, in which the item-response probabilities were constrained equal across
times and genders and the latent status prevalences and transition probabilities were
free to vary across genders, is shown in Table 8.6. Model 2 specified that all of
the parameters in the latent transition model were identical across genders. The
fit statistics for Models 1 and 2 and the hypothesis test appear in Table 8.7. The
difference G^2 was significant ($G_\Delta^2 = 111.8, df = 24, p < .0001$), and the AIC
suggested that Model 1 was preferred. However, the BIC suggested that the model in
which all parameters were equal across genders was preferred. We decided to reject
the null hypothesis for the time being, and to do some additional hypothesis testing
in an attempt to determine whether there were significant gender differences in latent
status prevalences, transition probabilities, or both. This additional hypothesis testing
is described in Sections 8.7.1 and 8.8.1.

Table 8.6   Latent Status Prevalences and Transition Probabilities for Five-Latent-Status Model of Adolescent Depression, by Gender (Add Health Public-Use Data, Waves I and II; $N = 2,061$)

	Latent Status				
	Not Depressed	Sad	Disliked	Sad + Disliked	Depressed
Latent status prevalences					
Males					
Time 1	.43	.18	.22	.08	.08
Time 2	.44	.19	.21	.08	.08
Females					
Time 1	.29	.27	.16	.14	.15
Time 2	.35	.27	.11	.14	.13
Males					
Probability of transitioning to...			*...Time 2 latent status*		
Conditional on...					
...Time 1 latent status					
Not Depressed	**.74**[†]	.11	.12	.00	.02
Sad	.27	**.57**	.09	.04	.02
Disliked	.29	.00	**.51**	.17	.03
Sad + Disliked	.03	.34	.26	**.28**	.09
Depressed	.10	.04	.04	.13	**.69**
Females					
Probability of transitioning to...			*...Time 2 latent status*		
Conditional on...					
...Time 1 latent status					
Not Depressed	**.78**	.13	.05	.04	.01
Sad	.32	**.52**	.02	.09	.06
Disliked	.19	.14	**.42**	.15	.11
Sad + Disliked	.03	.32	.12	**.40**	.13
Depressed	.08	.20	.04	.14	**.55**

$\ell = -14,478.8$.

[†]Diagonal transition probabilities in bold to facilitate interpretation.

Note. Item-response probabilities constrained equal across times.

Table 8.7   Fit Statistics for Test of Gender Differences in Latent Status Prevalences and Transition Probabilities for Adolescent Depression Example (Add Health Public-Use Data, Waves I and II; $N = 2,061$)

	G^2	df	AIC	BIC	ℓ
Model 1:					
Latent status prevalences and transition probabilities vary across gender	6,482.3	130,983	6,658.3	7,153.8	−14,478.8
Model 2:					
All parameters equal across gender	6,594.1	131,007	6,722.1	7,082.5	−14,534.7
$G_2^2 - G_1^2 = 111.8, df = 24, p < .0001$					

Note. Models 1 and 2 are described in Table 8.2.

8.7 TESTING THE HYPOTHESIS OF EQUALITY OF LATENT STATUS PREVALENCES ACROSS GROUPS AT TIME 1

Suppose that we are interested in testing the more specific hypothesis that the latent status prevalences at Time 1 are equal across groups. Table 8.2 suggests two different approaches to this hypothesis test. One approach is to compare Model 1, in which both the latent status prevalences at Time 1 and the transition probabilities are free to vary across groups, to Model 4, in which the transition probabilities remain free to vary but the latent status prevalences at Time 1 are constrained to be equal across groups. The other approach is to compare Model 2, in which both sets of parameters are constrained to be equal across groups, to Model 3, in which the latent status prevalences at Time 1 are free to vary across groups and the transition probabilities are constrained to be equal across groups.

How are these two approaches the same and how are they different? As was discussed above, both approaches compare models in which the item-response probabilities are constrained to be equal across times and groups. As Table 8.2 indicates, in each approach the comparison involves a pair of models that are identical, except that in one member of the pair the latent status prevalences are free to vary across groups (Models 1 and 3) and in the other member they are constrained to be equal across groups (Models 2 and 4). The difference between the two approaches lies in how the transition probabilities are treated. In the first approach, which compares Models 1 and 4, the transition probabilities are freely estimated. In the second approach, which compares Models 2 and 3, the transition probabilities are constrained to be equal across groups.

As long as the two models being compared are identical except that in one the latent status prevalences are constrained to be equal across groups and in the other the latent status prevalences can vary across groups, in other words, the two models are nested, both approaches are legitimate methods for testing the null hypothesis of invariance. However, the two approaches will not in general yield exactly the same results, although it would be unusual for them to produce dramatically different results. Neither of the two approaches may be considered superior to the other, but depending on the research questions at hand, one may be more appropriate conceptually than the other.

For example, suppose that the research question involves following change over time when there had been an initial random assignment to experimental conditions at Time 1. In this case it would probably make the most sense to compare Models 1 and 4, in which the transition probabilities are free to vary across groups. By contrast, if the hypothesis test is being done to follow up after an initial rejection of the overall hypothesis that both the latent status prevalences at Time 1 and the transition probabilities are equal across groups, it would probably make more sense to compare Models 2 and 3, in which the transition probabilities are constrained to be equal across groups.

8.7.1 Empirical example: Gender differences in adolescent depression

Let us return to the adolescent depression example. Table 8.8 shows the hypothesis tests associated with both approaches to testing the hypothesis of invariance of the latent status prevalences across genders. In both approaches the hypothesis test was significant ($G_4^2 - G_1^2 = 72.2, df = 4, p < .0001$ and $G_2^2 - G_3^2 = 79.8, df = 4, p < .0001$), and the AICs and BICs both suggested that the models in which the latent status prevalences can vary across genders were a better choice. Therefore, we reject the null hypothesis and conclude that the latent status prevalences are different across genders.

Figure 8.1 illustrates the prevalences of latent statuses of adolescent depression for males and females Table 8.6. As Figure 8.1 shows, compared to males at Time 1, females were less likely to be in the Not Depressed (.29 vs. .43) and Disliked latent statuses (.16 vs. .22) and more likely to be in the Sad (.27 vs. .18), Sad + Disliked (.14 vs. .08), and Depressed (.15 vs. .08) latent statuses.

Table 8.8 Fit Statistics for Two Approaches to Test of Gender Differences in Latent Status Prevalences for Adolescent Depression Example (Add Health Public-Use Data, Waves I and II; $N = 2,061$)

	G^2	df	AIC	BIC	ℓ
	Approach 1: Transition probabilities vary across gender				
Model 1: Latent status prevalences at Time 1 vary across gender	6,482.3	130,983	6,658.3	7,153.8	−14,478.8
Model 4: Latent status prevalences at Time 1 equal across gender	6,554.5	130,987	6,722.5	7,195.5	−14,514.9
$G_4^2 - G_1^2 = 72.2, df = 4, p < .0001$					
	Approach 2: Transition probabilities equal across gender				
Model 2: Latent status prevalences at Time 1 equal across gender	6,594.1	131,007	6,722.1	7,082.5	−14,534.7
Model 3: Latent status prevalences at Time 1 vary across gender	6,514.3	131,003	6,650.3	7,033.2	−14,494.8
$G_2^2 - G_3^2 = 79.8, df = 4, p < .0001$					

Note. Models 1 through 4 are described in Table 8.2.

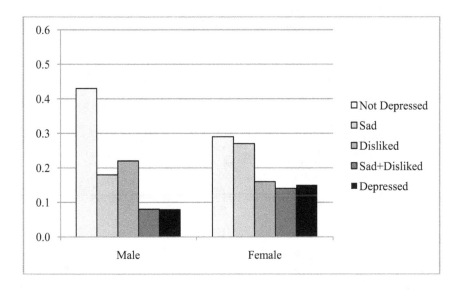

Figure 8.1   Prevalences of latent statuses of adolescent depression at Time 1 by gender (Add Health public-use data, Waves I and II; $N = 2,061$).

8.7.2   Empirical example: Gender differences in dating and sexual risk behavior

In Chapter 7 we presented a latent transition model of dating and sexual risk behavior that was originally reported in Lanza and Collins (2008). There were five latent statuses in this model: Nondaters, Daters, Monogamous, Multipartner Safe, and Multipartner Exposed.

Lanza and Collins (2008) also tested the hypothesis that the latent status prevalences at Time 1 were equal across genders. They allowed the transition probabilities to vary across genders while testing this hypothesis; in other words, they used Approach 1. Table 8.9 shows the fit statistics and hypothesis test associated with comparing Model 1, in which latent status prevalences and transition probabilities were allowed to vary across gender, with Model 4, in which the latent status prevalences at Time 1 were constrained to be equal across genders. In both Models 1 and 4, the item-response probabilities were constrained to be equal across groups. The hypothesis test was significant ($G_{\Delta}^2 = 71.5, df = 4, p < .0001$), and the AIC and BIC both suggested that Model 1 was preferred, so Lanza and Collins concluded that there were gender differences in Time 1 latent status prevalences.

Table 8.9 Fit Statistics for Test of Gender Differences in Latent Status Prevalences Across Gender for Dating and Sexual Risk Behavior Example (NLSY, Rounds 2–4; $N = 2,937$)

	G^2	df	AIC	BIC	ℓ
Model 1: Latent status prevalences at Time 1 vary across gender	3,422.1	93,193	3,658.1	4,364.4	−16,602.9
Model 4: Latent status prevalences at Time 1 equal across gender	3,493.6	93,197	3,721.6	4,404.0	−16,638.6
$G_4^2 - G_1^2 = 71.5, df = 4, p < .0001$					

Note. Models 1 and 4 are described in Table 8.2.

The parameter estimates for Model 1 appear in Table 8.10. As the table shows, compared to males at Time 1, females were more likely to belong to the Monogamous latent status (.18 versus .08) and less likely to belong to the Multi-Partner Safe latent status (.16 female versus .30 male). As might be expected, the prevalence of the Monogamous latent status increased consistently with time for both genders. However, the prevalence of the Monogamous latent status was considerably larger among females at every time. By contrast, whereas the prevalence of the Multipartner Safe latent status decreased with time for both groups, at every time this prevalence was smaller among females. The prevalence of the high-risk Multipartner Exposed latent status increased with time for both males and females, but the increase occurred at a faster rate for males.

8.8 TESTING THE HYPOTHESIS OF EQUALITY OF TRANSITION PROBABILITIES ACROSS GROUPS

In Section 8.7 we showed that there are two approaches to testing the hypothesis of equality of the latent status prevalences across groups. In much the same way, there are two approaches for testing the null hypothesis that the transition probabilities are equal across groups. One approach involves comparing Model 1, in which both sets of parameters are free to vary across groups, to Model 3, in which the latent status prevalences are free to vary and the transition probabilities are constrained to be equal across groups. The other approach involves comparing Model 2, in which both sets of parameters are constrained to be equal across groups, to Model 4, in which the latent status prevalences at Time 1 are constrained to be equal across groups and the transition probabilities are free to vary.

Table 8.10   Latent Status Prevalences and Transition Probabilities for Five-Latent-Status Model of Dating and Sexual Risk Behavior, by Gender (NLSY, Rounds 2–4; $N = 2{,}937$)

	Latent Status				
	Non-daters	Daters	Monogamous	Multi-partner Safe	Multi-partner Exposed
Latent status prevalences					
Males					
Time 1	.17	.28	.08	.30	.18
Time 2	.12	.24	.16	.26	.22
Time 3	.12	.17	.22	.22	.27
Females					
Time 1	.20	.30	.18	.16	.17
Time 2	.14	.24	.29	.15	.18
Time 3	.10	.20	.38	.10	.22
Males					
Probability of transitioning to...		...*time 2 latent status*			
Conditional on...					
...*time 1 latent status*					
Nondaters	**.61**[†]	.17	.09	.10	.03
Daters	.01	**.60**	.12	.22	.05
Monogamous	.03	.03	**.69**	.06	.18
Multipartner Safe	.04	.09	.14	**.58**	.14
Multipartner Exposed	.01	.07	.08	.00	**.84**
Probability of transitioning to...		...*time 3 latent status*			
Conditional on...					
...*time 2 latent status*					
Nondaters	**.74**	.11	.09	.06	.00
Daters	.06	**.48**	.16	.22	.09
Monogamous	.03	.04	**.62**	.02	.28
Multipartner Safe	.04	.11	.10	**.62**	.13
Multipartner Exposed	.02	.01	.21	.00	**.75**
Females					
Probability of transitioning to...		...*time 2 latent status*			
Conditional on...					
...*time 1 latent status*					
Nondaters	**.61**	.21	.08	.08	.03
Daters	.01	**.54**	.21	.18	.05
Monogamous	.05	.05	**.66**	.11	.13
Multipartner Safe	.04	.14	.35	**.40**	.08
Multipartner Exposed	.01	.00	.24	.00	**.75**
Probability of transitioning to...		...*time 3 latent status*			
Conditional on...					
...*time 2 latent status*					
Nondaters	**.53**	.21	.20	.05	.01
Daters	.01	**.58**	.26	.09	.06
Monogamous	.03	.06	**.69**	.00	.22
Multipartner Safe	.05	.08	.23	**.48**	.16
Multipartner Exposed	.01	.01	.32	.00	**.67**

$\ell = -16{,}602.9$.

[†] Diagonal transition probabilities in bold to facilitate interpretation.

Note. Item-response probabilities constrained equal across genders and times.

8.8.1   Empirical example: Gender differences in adolescent depression

Let us return to the adolescent depression example and use the two approaches to test the hypotheses that the transition probabilities are identical across genders. Results of application of the two approaches are shown in Table 8.11. Just as in the hypothesis tests described above, these two approaches to testing invariance are both legitimate but do not yield identical results.

Here the hypothesis tests and the information criteria point to different conclusions. Both hypothesis tests were significant ($G_3^2 - G_1^2 = 32.0, df = 20, p < .04$ and $G_2^2 - G_4^2 = 39.6, df = 20, p < .01$), suggesting that the null hypothesis that the transition probabilities are equal across genders should be rejected. However, the AICs and BICs suggested that the more parsimonious models that constrained the transition probabilities to be equal across genders were preferred.

Table 8.6 shows some evidence for gender differences in transition probabilities. For example, males who were in the Disliked latent statuses at Time 1 were more likely than females to be in the same latent status at Time 2 (.51 vs. .42), and males who were in the Depressed latent status were also more likely than females to be in the same latent status at Time 2 (.69 vs. .55). However, females who were in the Sad + Disliked latent status at Time 1 were more likely than males to be in that latent status at Time 2 (.40 vs. .28).

Table 8.11   Fit Statistics for Two Approaches to Test of Gender Differences in Transition Probabilities for Adolescent Depression Example (Add Health Public-Use Data, Waves I and II; $N = 2,061$)

	G^2	df	AIC	BIC	ℓ
	Approach 1: Latent status prevalences vary across genders				
Model 1: Transition probabilities vary across gender	6,482.3	130,983	6,658.3	7,153.8	−14,478.8
Model 3: Transition probabilities equal across gender $G_3^2 - G_1^2 = 32.0, df = 20, p < .04$	6,514.3	131,003	6,650.3	7,033.2	−14,494.8
	Approach 2: Latent status prevalences equal across genders				
Model 2: Transition probabilities equal across gender	6,594.1	131,007	6,722.1	7,082.5	−14,534.7
Model 4: Transition probabilities vary across gender $G_2^2 - G_4^2 = 39.6, df = 20, p < .01$	6,554.5	130,987	6,722.5	7,195.5	−14,514.9

Note. Models 1 through 4 are described in Table 8.2.

8.9   INCORPORATING COVARIATES IN LTA

The purpose of introducing covariates into a latent transition model is to identify characteristics that predict membership in the different latent statuses and/or predict transitions between latent statuses. As will be seen in the remainder of this chapter, the ideas that underlie the introduction of covariates in LTA are essentially the same as those that were introduced in Chapter 6 when we discussed adding covariates in LCA. Before reading this chapter the reader may wish to review Section 6.6, which explains how to interpret the estimates of intercepts and regression coefficients in LCA with covariates.

Just as we did in Chapter 6, here we strongly recommend that before introducing covariates into a latent transition model, the investigator should have a clear idea of what the latent structure is. For this reason, we recommend that before conducting a LTA with covariates, it is a good idea to go through the procedures described earlier in this book to select a baseline model without covariates. This model should provide an adequate representation of the data, have clearly interpretable latent statuses, and be identified. We also recommend incorporating any grouping variables into this baseline model and, where appropriate, assessing measurement invariance. Once this model has been arrived at, covariates may be introduced.

In Sections 8.11.1 and 8.11.3 we introduce covariates into the adolescent depression example as an empirical demonstration of predicting latent status membership at Time 1 and predicting transitions between latent statuses. We use three covariates: current cigarette use, lifetime marijuana use, and weekly alcohol use. As shown in these sections, interpreting the results of LTA in which covariates are used to predict latent status membership at Time 1 is fairly straightforward. However, interpreting the results of LTA in which covariates are used to predict transitions between latent statuses requires careful thought.

8.9.1   Missing data and preparing variables
for use as covariates

Missing data in covariates is handled the same way in LTA as it is in LCA (see Section 4.2.3). This means that when planning LTA with covariates, it can be helpful to think ahead about missing data. Some software for LTA with covariates allows missing data in the indicators of the latent classes but not in the covariates, and will automatically remove cases that have missing data on any covariate. If there are missing data on one or more covariates and no steps are taken to deal with this, the baseline model will be fit to a larger, and therefore somewhat different, data set than the covariate model. There are two strategies for avoiding this problem. One strategy is to remove any cases with missing data on any of the covariates before fitting the baseline model, so that the same data set is used in all analyses. However, if this results in a severe loss of data, issues of bias and a loss in statistical power may arise.

The other strategy is to use multiple imputation (Schafer, 1997; Schafer and Graham, 2002) to deal with missing data prior to any data analysis. This reduces or eliminates bias and preserves statistical power by enabling the investigator to make use of all available data. If multiple imputation is used, care must be taken to ensure that all covariates, along with any interactions between covariates, are present in the data set when the multiple imputation procedure is carried out.

Covariates and interactions between covariates are coded for use in LTA in the same manner as they are coded for use in LCA. There are two aspects of coding that we wish to note here. First, nominal covariates must be dummy-coded (transformed to variables with codes of 0 and 1) before being entered into the regression equation. Second, standardizing numeric covariates often makes interpretation of effects more straightforward. More about coding of covariates may be found in Section 6.3.1.

8.10   LTA WITH COVARIATES: MODEL AND NOTATION

In this section we use the adolescent depression example with current cigarette use as a covariate to illustrate the model for LTA with covariates.

As described in Section 8.3, suppose that there are $j = 1, ..., J$ observed indicator variables measuring the latent statuses, and the J observed variables have been measured at $t = 1, ..., T$ times. In the adolescent depression example, $J = 8$ and $T = 2$. Observed variable j has R_j response categories; in other words, to keep the exposition simple we assume that the number of response categories for observed variable j does not vary across times. In the example, $R_j = 2$ for all observed variables. At Time t, $r_{j,t} = 1, ..., R_j$ indexes the individual response categories.

Let L_t represent the categorical latent variable at Time t with S latent statuses, where $s_1 = 1, ..., S$ at Time 1, $s_2 = 1, ..., S$ at Time 2, and so on, up to $s_T = 1, ..., S$ at Time T. In the example the latent variable is adolescent depression, and there are $S = 5$ latent statuses at each time. In addition, there is a covariate X, current cigarette use. (To avoid making the notation overly complicated, we assume a single covariate.) This covariate is to be used to predict latent status membership at Time 1 and transitions between latent statuses at any two adjacent times. Finally, $I(y_{j,t} = r_{j,t})$ is an indictor function that equals 1 when the response to variable j at Time $t = r_{j,t}$, and equals 0 otherwise. (As mentioned previously, this function is merely a device for picking out the appropriate ρ parameters to multiply together.) Then the latent transition model can be expressed as follows:

$$P(\mathbf{Y} = \mathbf{y} | X = x)$$

$$= \sum_{s_1=1}^{S} \cdots \sum_{s_T=1}^{S} \delta_{s_1}(x) \tau_{s_2|s_1}(x) \ldots \tau_{s_T|s_{T-1}}(x) \prod_{t=1}^{T} \prod_{j=1}^{J} \prod_{r_{j,t}=1}^{R_j} \rho_{j,r_{j,t}|s_t}^{I(y_{j,t}=r_{j,t})}, \qquad (8.2)$$

where $\delta_{s_1}(x) = P(L_1 = s_1 | X = x)$ and $\tau_{s_t | s_{t-1}}(x) = P(L_t = s_t | L_{t-1} = s_{t-1}, X = x)$ are standard baseline-category multinomial logistic models (e.g., Agresti, 2007).

8.10.1 Predicting latent status membership

In most ways using covariates to predict latent status membership is essentially the same as using covariates to predict latent class membership. With a single covariate X, $\delta_{s_1}(x)$ can be expressed as follows:

$$\delta_{s_1}(x) = P(L_1 = s_1 | X = x) = \frac{e^{\beta_{0s_1} + \beta_{1s_1} x}}{1 + \sum_{s_1'=1}^{S-1} e^{\beta_0 s_1 + \beta_1 s_1 x}} \tag{8.3}$$

for $s_1' = 1, ..., S - 1$. Logistic regression requires designating one category of the criterion variable as the reference category. In this notation the reference category is the latent status prevalence corresponding to latent status S_1 (i.e., δ_{S_1}). As discussed in Chapter 6 (see Section 6.7.1.1), the choice of reference category is arbitrary and will not affect hypothesis testing, but it can affect the ease of interpretation of the results.

The logistic regression analysis produces an estimate of the effect for each latent status in comparison to the reference latent status. Thus, there will be $S-1$ regression coefficients β_{1s_1} corresponding to each covariate, plus $S - 1$ intercepts β_{0s_1}, for the latent status prevalences. For example, if there were three covariates, there would be three regression coefficients plus an intercept corresponding to each latent status except the one designated as the reference category.

8.10.2 Predicting transitions between latent statuses

Using covariates to predict transitions is a bit more complicated than using covariates to predict latent status membership at Time 1. Consider the $S \times S$ transition probability matrix

$$\begin{bmatrix} \tau_{1_t | 1_{t-1}} & \tau_{2_t | 1_{t-1}} & \cdots & \tau_{S_t | 1_{t-1}} \\ \tau_{1_t | 2_{t-1}} & \tau_{2_t | 2_{t-1}} & \cdots & \tau_{S_t | 2_{t-1}} \\ \cdots & \cdots & \cdots & \cdots \\ \tau_{1_t | S_{t-1}} & \tau_{2_t | S_{t-1}} & \cdots & \tau_{S_t | S_{t-1}} \end{bmatrix}. \tag{8.4}$$

Recall that each row of the transition probability matrix sums to 1 (Equation 7.4). When a covariate is used to predict transitions, there is a separate regression equation for each row of the transition probability matrix, and therefore each row of the transition probability matrix requires its own reference category. For example, consider the first row of the transition probability matrix in Equation 8.4, consisting of

$\tau_{1_t|1_{t-1}}, \tau_{2_t|1_{t-1}}, ..., \tau_{S_t|1_{t-1}}$. The elements in this row correspond to the probability of membership in latent status 1 at Time t, latent status 2 at Time t, and so on, among the subset of individuals who are in latent status 1 at the preceding time, Time $t - 1$. The logistic regression model for this row of the transition probability matrix enables the investigator to assess the effect of a covariate X on the transition probabilities for this subset of individuals.

So, for the first row of the transition probability matrix and with a single covariate X, $\tau_{s_t|1_{t-1}}(x)$ can be expressed as

$$\tau_{s_t|1_{t-1}}(x) = P(L_t = s_t|L_{t-1} = 1, X = x) = \frac{e^{\beta_{0s_t|1_{t-1}}+\beta_{1s_t|1_{t-1}}x}}{1+\sum_{s_t'}^{S-1} e^{\beta_{0s_t|1_{t-1}}+\beta_{1s_t|1_{t-1}}x}}$$

(8.5)

for $s_t' = 1, ..., S - 1$, where the reference category is latent status S. If there is one covariate, there will be $2(S - 1)$ regression coefficients for the first row of the transition probability matrix, in other words, an intercept $\beta_{0s_t|1_{t-1}}$ and slope $\beta_{1s_t|1_{t-1}}x$ for each latent status s_t except the latent status that has been designated as the reference category. This is repeated for each row of the transition probability matrix, corresponding to each of the latent statuses at Time $t - 1$. Thus, more generally,

$$\tau_{s_t|s_{t-1}}(x) = P(L_t = s_t|L_{t-1} = s_{t-1}, X = x) = \frac{e^{\beta_{0s_t|s_{t-1}}+\beta_{1s_t|s_{t-1}}x}}{1+\sum_{s_t'=1}^{S-1} e^{\beta_{0s_t|s_{t-1}}+\beta_{1s_t|s_{t-1}}x}}.$$

(8.6)

If there is one covariate, there will be $2S(S - 1)$ regression coefficients for the transition probabilities; that is, for each row of the transition probability matrix there is an intercept $\beta_{0s_t|s_{t-1}}$ and slope $\beta_{1s_t|s_{t-1}}x$ for each latent status s_t except the one that has been designated as the reference category.

We acknowledge that this notation is complex! As mentioned above, this is why we have presented these equations with a single covariate. However, multiple covariates can be incorporated. This is demonstrated in Sections 8.11.1 and 8.11.3.

8.10.3 Hypothetical example of LTA with covariates

To make the material presented in Sections 8.10.1 and 8.10.2 more concrete, let us review a hypothetical example. Suppose that a particular mathematics skill has been measured in children at two times, and three latent statuses emerge: Nonmasters, Intermediates, and Masters. Further suppose that grade-point average (GPA) is to be included as a covariate to predict mathematics skill latent status membership at Time 1 and latent status transitions between Times 1 and 2. Table 8.12 shows what regression equations would be involved and what intercepts and regression coefficients would be estimated. As the table shows, in this example there would be four regression equations, each of which estimates a set of intercepts and regression coefficients. The first regression equation uses GPA to predict latent status membership at Time 1. There is an intercept and a regression coefficient corresponding to each latent status except the Masters latent status, which has been (arbitrarily) designated as the reference category. The second, third, and fourth regression equations each pertain to a row of the transition probability matrix. Regression equation 2 uses GPA to predict latent status membership at Time 2 *for those who were in the Nonmasters latent status at Time 1*. Here, too, there is an intercept and a regression coefficient corresponding to each latent status except the reference latent status. Regression equation 3 uses GPA to predict latent status membership at Time 2 for those who were in the Intermediate latent status at Time 1, and Regression equation 4 uses GPA to predict latent status membership at Time 2 for those who were in the Masters latent status at Time 1.

8.10.4 What is estimated

Recall that in LCA with covariates, regression coefficients (β's) are estimated instead of latent class prevalences. In much the same way, in LTA with covariates the

Table 8.12 Hypothetical Example of Covariate (GPA) Predicting Latent Status Membership at Time 1 and Transitions Between Latent Statuses

	Latent Status		
	Non-masters (NM)	Intermediate (I)	Masters (M)
Regression equation 1:			
GPA predicting latent status membership at Time 1			
Intercepts	β_{0NM_1}	β_{0I_1}	ref
Regression coefficients	β_{1NM_1}	β_{1I_1}	ref
Regression equation 2:			
GPA predicting transitions to Time 2 latent status			
for those in the Nonmasters latent status at Time 1			
Intercepts	$\beta_{0NM_2|NM_1}$	$\beta_{0I_2|NM_1}$	ref
Regression coefficients	$\beta_{1NM_2|NM_1}$	$\beta_{1I_2|NM_1}$	ref
Regression equation 3:			
GPA predicting transitions to Time 2 latent status			
for those in the Intermediate latent status at Time 1			
Intercepts	$\beta_{0NM_2|I_1}$	$\beta_{0I_2|I_1}$	ref
Regression coefficients	$\beta_{1NM_2|I_1}$	$\beta_{1I_2|I_1}$	ref
Regression equation 4:			
GPA predicting transitions to Time 2 latent status			
for those in the Masters latent status at Time 1			
Intercepts	$\beta_{0NM_2|M_1}$	$\beta_{0I_2|M_1}$	ref
Regression coefficients	$\beta_{1NM_2|M_1}$	$\beta_{1I_2|M_1}$	ref

latent status prevalences and transition probabilities are not estimated when one or more covariates are included as predictors of those parameters. Instead, the regression coefficients (β_0's and β_1's) discussed above are estimated. The latent status prevalences and transition probabilities can be expressed as functions of the regression coefficients and individuals' values on the corresponding covariates. In both LCA and LTA with covariates, the item-response probabilities are estimated.

It is important to note that whereas in the model expressed in Equations 8.2, 8.3, 8.5, and 8.6 the covariate X can be related to the Time 1 latent status prevalences (i.e., the δ_{s_1}'s), and the transition probabilities (i.e., the τ's), they cannot be related to the item-response probabilities (i.e., the ρ's). This means that measurement invariance across all values of any covariates is assumed implicitly. (For an alternative approach, see Humphreys and Janson, 2000.) Note that measurement invariance across times is not required or assumed, although it can be imposed via parameter restrictions in the usual manner (see Section 7.11).

8.11 HYPOTHESIS TESTING IN LTA WITH COVARIATES

Hypothesis testing in LTA with covariates is very similar to hypothesis testing in LCA with covariates, described in Chapter 6. Hypothesis testing is performed by means of a likelihood-ratio χ^2 test. However, in LTA with covariates there are two possible hypothesis tests for each covariate. One hypothesis test concerns whether the covariate is a significant predictor of latent status membership at Time 1. The other hypothesis test concerns whether the covariate is a significant predictor of transitions between latent statuses.

Suppose that there is a single covariate X. Then the null hypothesis is tested by comparing the fit of Model 1, a baseline model without covariates and estimating p_1 parameters, to a corresponding Model 2, a model that includes the covariate X (as a predictor of latent status prevalences, transition probabilities, or both) and estimates p_2 parameters. Then $-2(\ell_1 - \ell_2)$ is in theory distributed as a χ^2 with $df = p_2 - p_1 =$ the number of regression coefficients associated with the covariate X.

In models with two or more covariates the null hypothesis is that the covariate of interest does not contribute significantly to prediction over and above the contribution of the other covariates in the model. This hypothesis can be tested by comparing the fit of a model that includes all of the covariates against that of a corresponding model that does not include the particular covariate being tested. Suppose that a latent transition model includes K covariates where $k = 1, ..., K$. The statistical significance of a single covariate X_k can be tested by means of a likelihood-ratio test that compares Model 1, a model that includes all covariates except X_k and estimates p_1 parameters, to Model 2, a model that includes all K covariates and estimates p_2 covariates. Then the hypothesis test proceeds in much the same way as described in the preceding paragraph; $-2(\ell_1 - \ell_2)$ is in theory distributed as a χ^2 with $df = p_2 - p_1$, where again $p_2 - p_1 = $ the number of regression coefficients associated with the covariate X_k.

If the significance of a set or block of covariates is to be tested, this can be accomplished by using essentially the same strategy, except that the difference between Models 1 and 2 involves the entire block of covariates. More detail on hypothesis tests involving covariates may be found in Section 6.5.

8.11.1 Empirical example of predicting latent status membership at Time 1: Adolescent depression

As mentioned above, we have chosen three covariates to add to the latent transition model of adolescent depression, all of which have to do with substance use. The covariates are current cigarette use, lifetime marijuana use, and weekly alcohol use. All three covariates were dummy coded so that 0 represented a response of "No" and 1 represented a response of "Yes."

The intercepts and regression coefficients are shown in Table 8.13. As the table indicates, the Not Depressed latent status served as the reference category. In addition to β's, Table 8.13 shows the odds associated with the intercept, and odds ratios associated with each of the covariates in the empirical example. As explained in Section 6.6, the intercepts (β_0's) in a logistic regression can be transformed into odds, and the regression coefficients (e.g., β_1's) can be transformed into odds ratios by exponentiating the coefficients (i.e., $e^{\beta_0} = $ odds and $e^{\beta_1} = $ odds ratios). Many people find odds and odds ratios to be more intuitively appealing and easier to interpret than β's.

8.11.1.1 *Interpreting the intercepts*

In general, the exponentiated intercepts (β_0's) reflect **the odds of membership in latent status s_1 in relation to the reference latent status S**. These odds can be computed using the latent status prevalences obtained from a latent transition model with no covariates. For example, Table 7.10 shows that the Time 1 prevalence of the reference latent status, Not Depressed, was $\delta = .36$, and that the Time 1 prevalence of the Disliked latent status was $\delta = .18$. The odds of membership in the Disliked latent status in relation to membership in the Not Depressed latent status were $.18/.36 = .49$ (within rounding error).

In this example the Not Depressed latent status has the largest prevalence at Time 1, so the odds of membership in each of the remaining latent statuses in relation to the Not Depressed latent status are all less than 1. Whenever the odds are less than 1, the corresponding β_0 is negative, as Table 8.13 shows.

Table 8.13   Substance-Use Predictors of Membership in Time 1 Latent Statuses of Adolescent Depression (Add Health Public-Use Data, Waves I and II; N = 2,061)

	Time 1 Latent Status				
	Not Depressed	Sad	Disliked	Sad + Disliked	Depressed
Intercept					
β_0's	ref	−.74	−.72	−1.41	−1.61
Odds	ref	.48	.49	.24	.20
Current Cigarette Use					
β_1's	ref	.57	.23	.44	.80
Odds Ratios	ref	1.78	1.26	1.56	2.23
Lifetime Marijuana Use					
β_2's	ref	.45	.04	.50	.49
Odds Ratios	ref	1.56	1.04	1.64	1.64
Weekly Alcohol Use					
β_3's	ref	−.25	−.30	−.33	−.02
Odds Ratios	ref	.78	.74	.72	.98

$\ell = -14,502.9$.

Note. All covariates entered simultaneously as predictors of Time 1 latent status membership.

8.11.1.2 Interpreting the regression coefficients

In general, the regression coefficients associated with prediction of latent status membership at Time 1 reflect **the change in odds of membership in latent status s_1 in relation to the reference latent status S associated with a one-unit increase in the predictor X.** A positive β_1 means that a one-unit increase in the covariate X is associated with increased odds. Therefore, the odds ratios associated with positive regression coefficients are always greater than 1. A negative β_1 means that a one-unit increase in the covariate X is associated with decreased odds. Therefore, the odds ratios associated with negative regression coefficients are always less than 1. When $\beta_1 = 0$, this means that a one-unit increase in the covariate is not associated with any change in odds. In this case the corresponding odds ratio is 1.

For example, Table 8.13 shows the β_1 and odds ratios associated with using the current cigarette use covariate to predict membership in the Depressed latent status at Time 1 in relation to membership in the Not Depressed latent status. $\beta_{1Depressed} = .80$ was positive, indicating that the odds of Time 1 membership in the Depressed latent status relative to the Not Depressed latent status were greater for those who reported current cigarette use than for those who reported no current cigarette use. This odds ratio was

$$odds\ ratio \;=\; \frac{\frac{P(Depressed|Current\ cigarette\ use=1)}{P(Not\ Depressed|Current\ cigarette\ use=1)}}{\frac{P(Depressed|Current\ cigarette\ use=0)}{P(NotDepressed|Current\ cigarette\ use=0)}}$$

$$= \quad e^{\beta_{1 Depressed}}$$

$$= \quad 2.23.$$

This can be interpreted in the following way: Suppose that there were two individuals. One individual reported current use of cigarettes and the other reported no current use of cigarettes. The individual who reported no current use of cigarettes had a particular odds of being in the Depressed latent status in relation to the Not Depressed latent status at Time 1; call these odds A. Similarly, the individual who reported current use of cigarettes had his or her own odds of being in the Depressed latent status in relation to the Not Depressed latent status at Time 1; call these odds B. The odds ratio of 2.23 means that these odds for the individual who reported current use of cigarettes were 2.23 times those of the individual who did not use cigarettes, that is, $B/A = 2.23$. In other words, this odds ratio means that at Time 1, an individual who reported current use of cigarettes was more than twice as likely to be in the Depressed latent status in relation to the Not Depressed latent status than was an individual who did not use cigarettes. Table 8.13 shows that the odds of membership in all of the latent statuses in relation to the Not Depressed latent status were greater for those who reported current cigarette use than for those who reported no current cigarette use.

Table 8.13 shows that the odds of membership in the Sad, Sad + Disliked, and Depressed latent statuses as compared to the Not Depressed latent status were greater for those who reported lifetime marijuana use than for those who reported no lifetime marijuana use. However, the odds of membership in the Disliked latent status relative to the Not Depressed latent status were approximately equal for those who had used marijuana and those who had not used marijuana. Figure 8.2 illustrates the odds ratios associated with current cigarette use and lifetime marijuana use.

8.11.1.3 Hypothesis tests

Table 8.14 shows the hypothesis test associated with each covariate. In each case $df = 4$. This is because for each covariate there were four regression coefficients, one corresponding to each latent status at Time 1, except the reference category. The degrees of freedom associated with each covariate used to predict latent status membership at Time 1 was always equal to the number of freely estimated latent status prevalences. As Table 8.14 shows, current cigarette use ($p = .0002$) and lifetime marijuana use ($p = .006$) were each related significantly to latent status membership at Time 1 when the other two covariates were in the model.

Weekly alcohol use ($p = .62$) was not a significant covariate when current cigarette use and lifetime marijuana use were in the model. This is reflected in the β's and odds ratios. Table 8.13 shows that all of the regression coefficients associated with this variable were close to 0, and all of the odds ratios were close to 1.

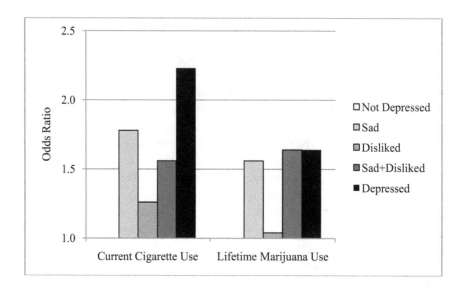

Figure 8.2   Odds ratios associated with current cigarette use and lifetime marijuana use. Although not estimated, the odds ratio of 1 associated with the reference latent status, Not Depressed, is shown here for comparison purposes (Add Health public-use data, Waves I and II; $N = 2,061$).

Table 8.14   Hypothesis Tests for Predictors of Membership in Latent Statuses of Adolescent Depression for Model Reported in Table 8.13

Covariate	ℓ Removing Covariate	Likelihood-Ratio Statistic	df	p
Current Cigarette Use	−14,514.1	22.4	4	.0002
Lifetime Marijuana Use	−14,510.2	14.4	4	.006
Weekly Alcohol Use	−14,504.3	2.6	4	.62

8.11.2   Empirical example of predicting latent status membership at Time 1: Dating and sexual risk behavior

As another empirical example, let us return to Lanza and Collins (2008). These authors introduced three covariates into their model of dating and sexual risk behavior and used them to predict latent status membership at Time 1. The three covariates were past-year cigarette use, past-year drunkenness, and past-year marijuana use. The cigarette and marijuana covariates were coded 0 if the individual reported no use of the substance during the past year and 1 if the individual reported any use of

Table 8.15   Substance-Use Predictors of Membership in Time 1 Latent Statuses of Dating and Sexual Risk Behavior (NLSY, Rounds 2–4; $N = 2,937$)

	Time 1 Latent Status				
	Non-daters	Daters	Monogamous	Multi-partner Safe	Multi-partner Exposed
Intercept					
β_0's	ref	.06	−1.14	−.43	−1.86
Odds	ref	1.06	.32	.65	.16
Past-Year Cigarette Use					
β_1's	ref	.69	1.03	1.25	1.15
Odds Ratios	ref	1.99	2.80	3.50	3.16
Past-Year Drunkenness					
β_2's	ref	1.21	1.30	1.26	2.12
Odds Ratios	ref	3.36	3.68	3.53	8.36
Past-Year Marijuana Use					
β_3's	ref	.52	.93	.95	2.36
Odds Ratios	ref	1.69	2.53	2.58	10.54

$\ell = -16,237.3$.

Note. All covariates entered simultaneously as predictors of Time 1 latent status membership.

the substance during the past year. The drunkenness covariate was coded 0 if the individual reported not having been drunk in the past year and 1 if the individual reported having been drunk in the past year.

Table 8.15 shows the intercepts, odds, regression coefficients, and odds ratios for the latent transition model with covariates. As the table shows, adolescents who reported using cigarettes or marijuana or having been drunk were roughly two to three times more likely than nonusers to be in the Daters or Monogamous latent status relative to the Nondaters latent status, and approximately three times more likely than nonusers to belong in the Multipartner Safe latent status relative to the Nondaters status. The effects of the different substances diverged substantially in prediction of the high-risk Multipartner Exposed latent status. Individuals who reported cigarette use in the past year were 3.16 times more likely than nonusers to be in the Multipartner Exposed latent status, whereas those who reported having been drunk or having used marijuana were 8.36 and 10.54 times more likely, respectively, to belong to this high-risk latent status relative to the Nondaters latent status.

Table 8.16 shows the hypothesis tests associated with the covariates in the model shown in Table 8.15. As the table shows, all three covariates were significant predictors of Time 1 latent status membership.

Table 8.16 Hypothesis Tests for Predictors of Membership in Latent Statuses of Dating and Sexual Risk Behavior for Model Reported in Table 8.15

Covariate	ℓ Removing Covariate	Likelihood-Ratio Statistic	df	p
Past-Year Cigarette Use	−16,278.5	82.5	4	<.0001
Past-Year Drunkenness	−16,277.8	81.0	4	<.0001
Past-Year Marijuana Use	−16,316.3	158.0	4	<.0001

8.11.3 Empirical example of predicting transitions between latent statuses: Adolescent depression

The three substance use covariates were introduced into the latent transition model of adolescent depression, this time as predictors of latent status transitions. The estimates of β parameters for prediction of latent status transitions in the adolescent depression example are shown in Table 8.17, and the corresponding odds ratios are shown in Table 8.18. Note that in every case the diagonal element of the transition probability matrix was the reference category. We chose to specify the diagonal element as the reference category so that all the odds ratios could be interpreted as the effect of the covariate on the odds of making a particular transition from Time 1 to Time 2 relative to being in the same latent status at Time 2.

8.11.3.1 Interpreting the regression coefficients

As mentioned above, interpreting the regression coefficients associated with prediction of latent status transitions requires careful thought. In general, the regression coefficients associated with prediction of latent status transitions from latent status s_t at Time t to latent status s_{t+1} at Time $t+1$ reflect **the change in odds of transitioning to latent status s_{t+1} in relation to the reference latent status S_{t+1}, conditional on membership in latent status s_t at Time t, associated with a one-unit increase in the predictor X.** As an example of how to interpret these estimates, consider the effect of current cigarette use on the transition from the Not Depressed latent status to the Depressed latent status. As Table 8.17 shows, the regression coefficient was .76, and as Table 8.18 shows, the odds ratio was 2.14. This odds ratio can be expressed as follows:

$$odds\ ratio = \frac{\frac{P(Depressed_2|Not\ Depressed_1, Current\ cigarette\ use=1)}{P(Not\ Depressed_2|Not\ Depressed_1, Current\ cigarette\ use=1)}}{\frac{P(Depressed_2|Not\ Depressed_1, Current\ cigarette\ use=0)}{P(Not\ Depressed_2|Not\ Depressed_1, Current\ cigarette\ use=0)}}$$

$$= e^{\beta_1 Depressed_2|Not\ Depressed_1}$$

Table 8.17   Logistic Regression Parameters (β's) for Substance-Use Predictors of Transitions Between Latent Statuses of Depression (Add Health Public-Use Data, Waves I and II; $N = 2,061$)

	Latent Status				
	Not Depressed	Sad	Disliked	Sad + Disliked	Depressed
Effect of current cigarette use on probability of transitioning to... Conditional on...			*...time 2 latent status*		
...time 1 latent status					
Not Depressed	ref	.32	−1.17	.70	.76
Sad	−.18	ref	.16	−1.82	.63
Disliked	.69	−1.01	ref	−.23	.70
Sad + Disliked	−.99	−.62	−1.07	ref	−.23
Depressed	1.19	−1.26	1.93	−1.04	ref
Effect of lifetime marijuana use on probability of transitioning to... Conditional on...			*...time 2 latent status*		
...time 1 latent status					
Not Depressed	ref	.69	−.38	.34	−3.93
Sad	−.27	ref	−.01	.23	−1.45
Disliked	−.73	−.33	ref	−.33	.38
Sad + Disliked	1.88	−.34	−.41	ref	−.26
Depressed	−.26	1.44	−2.32	.44	ref
Effect of weekly alcohol use on probability of transitioning to... Conditional on...			*...time 2 latent status*		
...time 1 latent status					
Not Depressed	ref	−.26	1.11	1.29	2.30
Sad	−.07	ref	−.42	−.65	.05
Disliked	.02	−.81	ref	−.87	−2.17
Sad + Disliked	1.72	.19	.34	ref	2.10
Depressed	−1.47	.51	−2.11	.46	ref

$\ell = -14,469.2$.

Note. All covariates entered simultaneously as predictors of Time 1 latent status membership and Time 1 – Time 2 transitions.

$$= \quad 2.14.$$

This can be interpreted in the following way: Suppose that there were two individuals, both of whom were in the Not Depressed latent status at Time 1. One of the individuals reported current use of cigarettes; the other reported no current use of cigarettes. The individual who reported no current use of cigarettes had a particular odds of being in the Depressed latent status at Time 2 in relation to being in the Not Depressed latent status at Time 2; call these odds A. Similarly, the individual who reported current use of cigarettes had his or her own odds of being in the Depressed latent status at Time 2 in relation to being in the Not Depressed latent status at Time 2; call these odds B. The odds ratio of 2.14 means that these odds for the individual who reported current

Table 8.18   Odds Ratios for Substance-Use Predictors of Transitions Between Latent Statuses of Depression (Add Health Public-Use Data, Waves I and II; $N = 2,061$)

	Latent Status				
	Not Depressed	Sad	Disliked	Sad + Disliked	Depressed
Effect of current cigarette use on probability of transitioning to... Conditional on... ...time 1 latent status		...*time 2 latent status*			
Not Depressed	ref	1.37	.31	2.01	2.14
Sad	.83	ref	1.17	.16	1.87
Disliked	1.99	.37	ref	.79	2.01
Sad + Disliked	.37	.54	.34	ref	.79
Depressed	3.29	.28	6.87	.35	ref
Effect of lifetime marijuana use on probability of transitioning to... Conditional on... ...time 1 latent status		...*time 2 latent status*			
Not Depressed	ref	2.00	.69	1.40	.02
Sad	.76	ref	.99	1.26	.23
Disliked	.48	.72	ref	.72	1.46
Sad + Disliked	6.58	.71	.66	ref	.77
Depressed	.77	4.21	.10	1.55	ref
Effect of weekly alcohol use on probability of transitioning to... Conditional on... ...time 1 latent status		...*time 2 latent status*			
Not Depressed	ref	.77	3.02	3.64	9.93
Sad	.93	ref	.66	.52	1.05
Disliked	1.02	.45	ref	.42	.11
Sad + Disliked	5.58	1.20	1.41	ref	8.17
Depressed	.23	1.67	.12	1.58	ref

$\ell = -14,469.2$.

Note. All covariates entered simultaneously as predictors of Time 1 latent status membership and Time 1 to Time 2 transitions.

use of cigarettes were 2.14 times greater than those of the individual who reported no current use of cigarettes; that is, $B/A = 2.14$.

In other words, this odds ratio means that among individuals who started out in the Not Depressed latent status at Time 1, those who reported current use of cigarettes were more than twice as likely to be in the Depressed latent status in relation to the Not Depressed latent status at Time 2 than were those who did not currently use cigarettes.

As another example, consider the effect of lifetime marijuana use on the transition from the Sad latent status to the Disliked latent status. Table 8.17 shows that the regression coefficient was −.01, and Table 8.18 shows that the odds ratio was .99.

This odds ratio can be expressed as follows:

$$odds\ ratio \quad = \quad \frac{\frac{P(Disliked_2|Sad_1,Lifetime\ marijuana\ use=1)}{P(Sad_2|Sad_1,Lifetime\ marijuana\ use=1)}}{\frac{P(Disliked_2|Sad_1,Lifetime\ marijuana\ use=0)}{P(Sad_2|Sad_1,Lifetime\ marijuana\ use=0)}}$$

$$= \quad e^{\beta_1 Disliked_2|Sad_1}$$

$$= \quad .99.$$

This can be interpreted in the following way. Suppose that there were two individuals, both of whom were in the Sad latent status at Time 1. One of the individuals reported having used marijuana; the other reported not having used marijuana. The individual who reported not having used marijuana had a particular odds of being in the Disliked latent status at Time 2 in relation to being in the Sad latent status at Time 2; call these odds A. Similarly, the individual who reported current use of marijuana had his or her own odds of being in the Disliked latent status at Time 2 in relation to being in the Sad latent status at Time 2; call these odds B. The odds ratio of .99 means that these odds for the individual who reported current use of cigarettes were about the same as those of the individual who reported no current use of marijuana; that is, $B/A = .99$.

In other words, this odds ratio means that among individuals who started out in the Sad latent status at Time 1, those who reported having used marijuana were about as likely to be in the Disliked latent status in relation to the Sad latent status at Time 2 as those who did not report having used marijuana.

8.11.3.2 Hypothesis tests

Table 8.19 shows the hypothesis test associated with each covariate. In each case $df = 20$. Recall from Section 8.10 that there is a separate regression for each row of the transition probability matrix. Thus, in this example there were five separate regression equations for predicting transitions. Associated with each of these regression equations there were four regression coefficients for each covariate, that is, one corresponding to each transition except the one corresponding to the reference category. Thus, $df = 5 \times 4 = 20$. The degrees of freedom associated with each covariate used to predict transitions between latent statuses is always equal to the number of freely estimated transition probabilities.

As Table 8.19 shows, none of the three substance-use variables were significant predictors of transitions between latent statuses of depression. This may seem surprising. Looking at Tables 8.17 and 8.18 overall, it is evident that there were several β's that were large in absolute value and correspondingly, several odds ratios that deviated considerably from 1. For example, we discussed above how for individuals who started out in the Not Depressed latent status, those who reported use of cigarettes were more than twice as likely to be in the Depressed latent status at Time

Table 8.19   Hypothesis Tests for Predictors of Transitions Between Latent Statuses of Adolescent Depression for Model Reported in Tables 8.17 and 8.18

Covariate	ℓ Removing Covariate	Likelihood-Ratio Statistic	df	p
Current Cigarette Use	−14,480.2	22.0	20	.34
Lifetime Marijuana Use	−14,479.4	20.4	20	.43
Weekly Alcohol Use	−14,479.5	20.6	20	.42

2 in relation to the reference latent status. Although this odds ratio would seem to suggest that current cigarette use was associated with heightened risk for making this particular transition, Table 8.19 shows that this relation was not significant.

It is not uncommon for substantial β's and odds ratios associated with prediction of transitions between latent statuses not to reach statistical significance. It is important to remember that each row of the transition probability matrix is based on only that part of the sample that were in the row latent status at Time t. Thus some transition probabilities may be associated with small sample sizes. This in turn can cause low statistical power.

The small sample sizes in individual rows of the transition probability matrix can also lead to instability in parameter estimation. Sometimes estimation can even break down if the sample size for a particular transition is very small. Binomial logistic regression, discussed in Section 8.13, is one method for dealing with this estimation problem.

8.11.4   Empirical example of predicting transitions between latent statuses: Dating and sexual risk behavior

Lanza and Collins (2008) used the binary variable reflecting whether an individual reported having been drunk in the past year, described in Section 8.11.2, as a covariate predicting latent status transitions from Time 1 to Time 2. Table 8.20 shows the regression coefficients and odds ratios associated with each latent status transition relative to staying in the same latent status over time (note that the three transition probabilities that were fixed to zero are not included in the multinomial regression equation for that row). Drunkenness appeared to be associated with an increased probability of transitioning from the Nondaters and Daters latent statuses to the Multipartner Exposed latent status relative to remaining in the same low-risk latent status over time (odds ratio = 3.59 for Nondaters; odds ratio = 2.98 for Daters). Also, those who had engaged in drunkenness in the past year appeared to be less likely to transition from the high-risk Multipartner Exposed status to the Monogamous status (odds ratio = .69). However, the drunkenness covariate was not found to be a

Table 8.20 Logistic Regression Parameters (β's) and Odds Ratios for Drunkenness Predicting Transitions Between Latent Statuses of Dating and Sexual Risk Behavior (NLSY, Rounds 2–4; $N = 2,937$)

	Latent Status				
	Non-daters	Daters	Monogamous	Multi-partner Safe	Multi-partner Exposed
β's					
Effect of past-year drunkenness on probability of transitioning to...			...*time 2 latent status*		
Conditional on...					
...*time 1 latent status*					
Nondaters	ref	−.09	.46	.09	1.28
Daters	−.52	ref	.67	.49	1.09
Monogamous	−.12	−.02	ref	$0^\dagger$	−.03
Multipartner Safe	$0^\ddagger$	$0^\ddagger$	$0^\ddagger$	$0^\ddagger$	$0^\ddagger$
Multipartner Exposed	$0^\dagger$	.26	−.36	$0^\dagger$	ref
Odds ratios					
Effect of past-year drunkenness on probability of transitioning to...			...*time 2 latent status*		
Conditional on...					
...*time 1 latent status*					
Nondaters	ref	.91	1.58	1.10	3.59
Daters	.60	ref	1.96	1.63	2.98
Monogamous	.89	.98	ref	$1.0^\dagger$	.97
Multipartner Safe	$1.0^\ddagger$	$1.0^\ddagger$	$1.0^\ddagger$	$1.0^\ddagger$	$1.0^\ddagger$
Multipartner Exposed	$1.0^\dagger$	1.30	.69	$1.0^\dagger$	ref

$\ell = -11,281.3$.

† Transition probability was trivially small and therefore fixed to zero. This latent status not included in logistic regression for that row of matrix.

‡ For at least one level of the covariate, one cell in row of transition probability matrix was empty. Logistic regression not conducted for this row.

Note. Covariate entered as predictor of Time 1 latent status membership and transitions between latent statuses.

significant predictor of transitions from Time 1 to Time 2 in dating and sexual risk behavior.

8.12 INCLUDING BOTH A GROUPING VARIABLE AND A COVARIATE IN LTA

Like latent class models, latent transition models can include both a grouping variable and a covariate. This approach enables the investigator to examine whether there are group differences in the regression coefficients and odds ratios associated with each covariate.

8.12.1   Empirical example:
Dating and sexual risk behavior

An example of a latent transition model with both a grouping variable and covariates appears in Lanza and Collins (2008). They used gender as a grouping variable to examine the differential effects of substance use predictors on Time 1 membership in dating and sexual risk behavior latent statuses. The results are shown in Table 8.21. Overall, the pattern looks similar for males and females, with a few exceptions. For instance, females who reported engaging in cigarette use were more likely than males to belong to the Multipartner Safe (odds ratios of 6.78 vs. 2.40) latent status at Time 1 relative to the Nondaters latent status. Females who reported drunkenness in the past year were less likely than males to belong to the Monogamous latent status (odds ratios of 3.20 vs. 5.46) relative to the Nondaters latent status.

Note that a formal hypothesis test of the significance of gender as a moderator of the effects of substance use behavior can be obtained by fitting a model with gender, substance use, and the gender × substance use interaction included as covariates, rather than including gender as a grouping variable. The test of the significance of the gender × substance use interaction provides a test of the significance of gender as a moderator of the effects of substance use behavior.

8.13   COMBINING LATENT STATUSES TO APPLY
BINOMIAL LOGISTIC REGRESSION IN LTA

In Section 6.12 we described a binomial logistic regression approach to LCA with covariates. In this approach, the investigator selects a particular latent class as the *target* latent class, and then combines the remaining latent classes into a single reference latent class, resulting in an outcome with only two categories, irrespective of the original number of latent classes. This idea can be directly extended to LTA. The approach may be used in prediction of Time 1 latent status membership, latent status transitions, or both.

The reasons for taking a binomial approach in LTA with covariates are similar to the reasons underlying this decision in LCA. One reason for using binomial logistic regression is to highlight a particular research question. Another reason is to deal with estimation difficulties stemming from the presence of small latent statuses. In LTA such estimation problems are particularly common in prediction of latent status transitions. As discussed in Section 8.11.3.2, each row of a transition probability matrix expressing transitions between Time t and Time $t + 1$ is conditional on membership in a particular latent status at Time t. If the prevalence of that latent status at Time t is small, then some or all of the transition probabilities in the row may be based on very small N's, which in turn can lead to difficulties in estimating regression coefficients when covariates are introduced. These estimation problems

Table 8.21   Substance-Use Predictors of Membership in Time 1 Latent Statuses of Dating and Sexual Risk Behavior by Gender (NLSY, Rounds 2–4; $N = 2,937$)

	Time 1 Latent Status				
	Non-daters	Daters	Monogamous	Multi-partner Safe	Multi-partner Exposed
Effect for Males					
Past-Year Cigarette Use					
β_1's	ref	.45	1.19	.88	1.13
Odds Ratios	ref	1.57	3.30	2.40	3.10
Past-Year Drunkenness					
β_2's	ref	1.31	1.70	1.23	2.08
Odds Ratios	ref	3.72	5.46	3.42	8.01
Past-Year Marijuana Use					
β_3's	ref	.47	.86	.91	2.39
Odds Ratios	ref	1.61	2.37	2.49	10.90
Effect for females					
Past-Year Cigarette Use					
β_1's	ref	.93	.98	1.91	1.16
Odds Ratios	ref	2.54	2.67	6.78	3.18
Past-Year Drunkenness					
β_2's	ref	1.06	1.16	1.17	2.19
Odds Ratios	ref	2.89	3.20	3.22	8.93
Past-Year Marijuana Use					
β_3's	ref	.56	1.16	.78	2.34
Odds Ratios	ref	1.75	3.20	2.19	10.40

$\ell = -16,136.1$.

Note. All covariates entered simultaneously as predictors of Time 1 latent status membership.

can often be eliminated by combining latent statuses and taking a binomial approach. In Section 8.13.1 we demonstrate the use of binomial logistic regression in prediction of latent status transitions.

In LTA with covariates, there is a regression equation corresponding to the latent status prevalences at Time 1, plus a regression equation corresponding to each row of the transition probability matrix. For each of these binomial logistic regression equations, one latent status is selected as the target and the remaining $S - 1$ latent statuses are combined into a single reference latent status. For each regression this creates an outcome variable with two categories. Different target latent statuses may be chosen for each regression equation.

8.13.1   Empirical example: Adolescent depression

Suppose an investigator is particularly interested in predicting transitions into the Depressed latent status in the adolescent depression example. From this perspective, it might be clearer to distinguish between just two types of transitions: those in which the individual ends up in the Depressed latent status at Time 2, and those in which the individual ends up in any other latent status at Time 2. In this case combining latent statuses and applying binomial logistic regression might be preferable.

In this example, the Depressed latent status was the target latent status in each row of the transition probability matrix. The four remaining latent statuses in each row were combined to form the reference latent status for that row. In other words, the latent transition model still comprised five latent statuses of depression, but the effects of the covariates reflected any increase in odds of membership in the Depressed latent status at Time 2 relative to membership in any of the other four latent statuses at Time 2.

Estimates of regression coefficients and odds ratios for the transitions into the Depressed latent status appear in Table 8.22. The largest odds ratio, 11.71, was associated with weekly alcohol use as a predictor of membership in the Depressed latent status at Time 2 relative to membership in any other latent status at Time 2 for individuals in the Not Depressed latent status at Time 1. This odds ratio means that among individuals who were in the Not Depressed latent status at Time 1, those who reported weekly use of alcohol were nearly 12 times as likely to be in the Depressed latent status at Time 2 relative to the Not Depressed latent status than were the individuals who did not use alcohol weekly. It is interesting to note that this analysis implicitly controls for depression at Time 1, because each row of the transition probability matrix is conditioned on Time 1 depression latent status.

8.13.1.1 *Hypothesis tests*

Table 8.23 shows hypothesis tests for a latent transition model in which multinomial logistic regression was used to predict adolescents' membership in depression latent statuses at Time 1 and binomial logistic regression was used to predict transitions into the Depressed latent status between Times 1 and 2. As would be expected, the hypothesis tests associated with predicting Time 1 depression had the same degrees of freedom as those in Table 8.14, and the tests themselves were nearly identical. The tests associated with prediction of latent status transitions were different from their counterparts in Table 8.19. The hypothesis tests in Table 8.19 were associated with a total of 20 degrees of freedom because four regression coefficients were estimated for each row of the transition probability matrix. By contrast, the hypothesis tests for prediction of the transition into Depression relative to any other latent status were associated with a total of 5 degrees of freedom, because one regression coefficient was estimated for each row of the transition probability matrix. In the multinomial logistic regressions there was a parameter corresponding to each element of the transition probability matrix except the reference category in each row. This held for the binomial logistic regression as well, but in this model there were only two categories in each row, the reference category and the target category, so there was one parameter corresponding to each row for a total of $df = 5$.

Table 8.22   Effect of Substance-Use Predictors on Probability of Transition to
Depressed Latent Status Relative to All Other Latent Statuses Conditional on Time 1
Latent Status (Add Health Public-Use Data, Waves I and II; N = 2,061)

Time 1 Latent Status	β	Odds Ratio
Effect of current cigarette use		
Not Depressed	1.04	2.83
Sad	.49	1.64
Disliked	.60	1.82
Sad + Disliked	.58	1.79
Depressed	.20	1.23
Effect of lifetime marijuana use		
Not Depressed	−3.79	.02
Sad	−1.54	.22
Disliked	.63	1.88
Sad + Disliked	−.21	.81
Depressed	−.40	.67
Effect of weekly alcohol use		
Not Depressed	2.46	11.71
Sad	.24	1.28
Disliked	−2.08	.13
Sad + Disliked	1.75	5.77
Depressed	.08	1.08

$\ell = -14{,}492.7$.

Note. All covariates entered in one model as predictors of Time 1 latent status membership (using multinomial logistic regression) and transition probabilities (using binomial logistic regression).

As Table 8.23 shows, current cigarette use ($p = .76$) and lifetime marijuana use ($p = .16$) were not significant predictors of the transition into the Depressed latent status. Weekly alcohol use was marginally significant ($p = .08$).

8.13.2   Empirical example: Dating and sexual risk behavior

In their latent transition analysis of dating and sexual risk behavior, Lanza and Collins (2008) were particularly interested in using the past-year drunkenness covariate to predict transitions into the latent status representing the highest risk, namely Multipartner Exposed. Therefore, they conducted a binomial logistic regression. In this regression, for each row of the transition probability matrix, the reference group was the first four latent statuses collapsed together, and the target group was Multipartner Exposed. Results for this binomial logistic regression appear in Table 8.24. As the table shows, for each Time 1 latent status of dating and sexual risk behavior, past-year drunkenness was associated with increased odds of membership in the Multipartner Exposed latent status at Time 2 relative to membership in all other latent statuses combined. These effects were strongest for individuals who initially

Table 8.23 Hypothesis Tests for Predictors of Membership in Latent Statuses of Adolescent Depression at Time 1 and Transition to Depressed Latent Status for Model Reported in Table 8.22

Covariate	ℓ Removing Covariate	Likelihood-Ratio Statistic	df	p
Prediction of Time 1 membership in Depressed latent status				
Current Cigarette Use	−14,503.4	21.5	4	.0003
Lifetime Marijuana Use	−14,500.0	14.7	4	.005
Weekly Alcohol Use	−14,493.8	2.3	4	.68
Prediction of transition to Depressed latent status				
Current Cigarette Use	−14,494.0	2.6	5	.76
Lifetime Marijuana Use	−14,496.7	8.0	5	.16
Weekly Alcohol Use	−14,497.6	9.8	5	.08

Note. Prediction of Time 1 latent statuses based on multinomial logistic regression; prediction of transition to Depressed latent status based on binomial logistic regression. For each covariate one β is estimated for each of five rows in the transition probability matrix.

were Nondaters or Daters. The effect of the past-year drunkenness covariate was significant (likelihood-ratio statistic = 14.0, df = 4, p = .007).

Table 8.24 Effect of Past-Year Drunkenness on Probability of Transition to Multipartner Exposed Latent Status at Time 2 Conditional on Time 1 Latent Status (NLSY, Rounds 2–4; N = 2,937)

Time 1 Latent Status	β	Odds Ratio
Nondaters	1.36	3.90
Daters	1.19	3.29
Monogamous	.44	1.56
Multipartner Safe	0[†]	1.0[†]
Multipartner Exposed	.46	1.59

ℓ = −11,282.7.

[†]For at least one level of the covariate, either the reference or target group in this row of the transition probability matrix was empty. Logistic regression not conducted for this row.

Note. Covariate entered as predictor of Time 1 latent status membership using multinomial model and transition probabilities using binomial model.

8.14 THE RELATION BETWEEN MULTIPLE-GROUP LTA AND LTA WITH A COVARIATE

In Chapter 6 we explained how LCA with a grouping variable and LCA with covariates are related. Essentially the same relation holds between LTA with a grouping variable and LTA with covariates.

A latent transition model with a grouping variable and with the item-response probabilities constrained to be equal across groups is mathematically equivalent to a particular latent transition model with covariates. The equivalent covariate model is identical to the grouping variable model except that the grouping variable has been transformed into one or more dummy-coded variables and these variables have been included in the LTA as covariates.

Similarly, LTA with both a grouping variable and a covariate X, and with the item-response probabilities constrained to be equal across groups, is mathematically equivalent to another latent transition model. The equivalent covariate model is identical to the grouping variable model except that the grouping variable has been transformed into one or more dummy-coded variables, and these variables, the co-variate X, and the interactions between the dummy-coded variables and the covariate X appear in the model as predictors. A more complete discussion of when LCA with grouping variables and LCA with covariates are equivalent appears in Section 6.10.2.

Note that this equivalence does not hold in general when latent statuses have been combined in order to apply binomial logistic regression (see Section 6.12.2).

8.15 SUGGESTED SUPPLEMENTAL READINGS

Lanza and Collins (2008) includes examples of both multiple-group LTA and LTA with covariates for the model of transitions in dating and sexual risk behavior discussed in the current chapter. The article also explains how to fit these models using Proc LTA. Collins, Graham, Rousculp, and Hansen (1997) compared the substance use onset process for children who had and had not exhibited heavy use of caffeine. Guo, Collins, Hill, and Hawkins (2000) compared alcohol use onset in middle and high school for those who later were and were not diagnosed with alcohol abuse or dependence. Lanza and Collins (2002) compared the substance use onset process for girls who experienced puberty early as compared to those who experienced puberty on time or late, using a multiple-group approach. Chung, Park, and Lanza (2005) followed up on these analyses using age as a covariate. Maldonado-Molina, Collins, Lanza, Prado, Ramírez, and Canino (2007) used LTA to compare patterns of substance use onset in Hispanic youth living in the United States to those of youth living in Puerto Rico.

8.16   POINTS TO REMEMBER

• Before conducting a multiple-group LCA or a LCA with covariates it is a good idea to establish a baseline model without grouping variables or covariates.

• Multiple-group LTA is very similar to multiple-group LCA. In multiple-group LTA, estimates of latent status prevalences at Time 1, transition probabilities, and item-response probabilities can be conditioned on group.

• Multiple-group LTA can be used to examine whether measurement invariance holds across groups. Measurement invariance may hold across both times and groups, across groups but not times, across times but not groups, or across neither times nor groups.

• Multiple-group LTA can be used to examine whether latent status prevalences at Time 1 are equivalent across groups, and also whether transition probabilities are equivalent across groups. There are two equally valid, but not identical, approaches for conducting these hypothesis tests.

• LTA with covariates is very similar to LCA with covariates. Like LCA with covariates, in LTA with covariates prediction is done using a logistic regression approach.

• Covariates in LTA may be used to predict latent status membership at Time 1 or transitions between latent statuses.

• The model for LTA with covariates shown in this book assumes measurement invariance across all levels of covariates. In this model item-response probabilities cannot be modeled as functions of covariates.

• Covariates predict latent status membership at Time 1 and transitions between latent statuses separately. There is one regression equation corresponding to prediction of latent status membership at Time 1. In addition, corresponding to prediction of transitions between latent statuses there is a regression equation for each row of each transition probability matrix.

• It is necessary to choose one latent status as a reference category for each regression equation. A different reference latent status can be chosen for each equation. All odds and odds ratios are computed in relation to the reference latent status. This choice does not affect hypothesis testing, but it can have an impact on how readily results can be interpreted.

• Odds ratios corresponding to prediction of transitions between latent statuses are complex, and interpretation of the regression coefficients requires careful thought.

• Binomial logistic regression can be used to predict odds of membership in a target latent status as compared to a reference category that is an aggregate of all remaining latent statuses. Binomial logistic regression may be used to predict latent status membership at Time 1, transitions between latent statuses, or both. This approach may be particularly useful in prediction of transitions between latent statuses, because it can deal with certain common estimation problems.

Bibliography

A. Agresti. *Categorical data analysis*. Wiley, New York, 1990.

A. Agresti. *An introduction to categorical data analysis (Second edition)*. Wiley, Hoboken, NJ, 2007.

L.S. Aiken and S.G. West. *Multiple regression: Testing and interpreting interactions*. Sage, Thousand Oaks, CA, 1991.

M. Aitkin, D. Anderson, and J. Hinde. Statistical modelling of data on teaching styles. *Journal of the Royal Statistical Society. Series A*, 144:419–461, 1981.

H. Akaike. Factor analysis and AIC. *Psychometrika*, 52:317–332, 1987.

T. Asparouhov and B.O. Muthén. Multilevel mixture models. In G. R. Hancock and K. M. Samuelsen, editors, *Advances in latent variable mixture models*, pages 27–51. Information Age Publishing, Charlotte, NC, 2008.

K.J. Auerbach and L.M. Collins. A multidimensional developmental model of alcohol use during emerging adulthood. *Journal of Studies on Alcohol*, 67:917–925, 2006.

F.B. Baker and S.-H. Kim. *Item response theory: Parameter estimation techniques (Second edition)*. Marcel Dekker, New York, NY, 2004.

K. Bandeen-Roche, D.L. Miglioretti, S.L. Zeger, and P.J. Rathouz. Latent Variable Regression for Multiple Discrete Outcomes. *Journal of the American Statistical Association*, 92:1375–1386, 1997.

L.R. Bergman and D. Magnusson. A person-oriented approach in research on developmental psychopathology. *Development and Psychopathology*, 9:291–319, 1997.

L.R. Bergman, D. Magnusson, and B.M. El-Khouri. *Studying individual development in an interindividual context: A person-oriented approach*. Lawrence Erlbaum Associates, Mahwah, NJ, 2003.

U. Böckenholt. Mixed-effects analyses of rank-ordered data. *Psychometrika*, 66: 45–62, 2001.

K.A. Bollen. *Structural equations with latent variables*. Wiley, New York, 1989.

K.A. Bollen. Latent variables in psychology and the social sciences. *Annual Review of Psychology*, 53:605–634, 2002.

K.A. Bollen and P.J. Curran. *Latent curve models: A structural equation perspective*. Wiley, Hoboken, NJ, 2005.

G.E.P. Box and G.M. Jenkins. *Time series analysis: Forecasting and control, rev. ed.* Holden-Day, San Francisco, 1976.

H. Bozdogan. Model selection and Akaike's Information Criterion (AIC): The general theory and its analytical extensions. *Psychometrika*, 52:345–370, 1987.

C.M. Bulik, P.F. Sullivan, and K.S. Kendler. An empirical study of the classification of eating disorders. *American Journal of Psychiatry*, 157:886–895, 2000.

B.V. Bye and E.S. Schechter. A latent Markov model approach to the estimation of response error in multiwave panel data. *Journal of the American Statistical Association*, 81:375–380, 1986.

G. Celeux and G. Soromenho. An entropy criterion for assessing the number of clusters in a mixture model. *Journal of Classification*, 13:195–212, 1996.

Centers for Disease Control and Prevention. Methodology of the Youth Risk Behavior Surveillance System. *Morbidity and Mortality Weekly Report*, 53:1–13, 2004.

Centers for Disease Control and Prevention. Youth Risk Behavior Survey. http://www.cdc.gov/yrbss, January 2005.

H. Chung, E. Loken, and J.L. Schafer. Difficulties in drawing inferences with finite-mixture models: A simple example with a simple solution. *The American Statistician*, 58:152–159, 2004.

H. Chung, Y.S. Park, and S.T. Lanza. Latent transition analysis with covariates: Pubertal timing and substance use behaviours in adolescent females. *Statistics in Medicine*, 24:2895–2910, 2005.

H. Chung, B.P. Flaherty, and J.L. Schafer. Latent class logistic regression: Application to marijuana use and attitudes among high school seniors. *Journal of the Royal Statistical Society: Series A*, 169:723–743, 2006.

H. Chung, S.T. Lanza, and E. Loken. Latent transition analysis: Inference and estimation. *Statistics in Medicine*, 27:1834–1854, 2008.

T. Chung and C.S. Martin. Classification and short-term course of DSM-IV cannabis, hallucinogen, cocaine, and opioid disorders in treated adolescents. *Journal of Consulting and Clinical Psychology*, 73:995–1004, 2005.

M.J. Cleveland, L.M. Collins, S.T. Lanza, M.T. Greenberg, and M.E. Feinberg. Does individual risk moderate the effect of contextual-level protective factors? A latent class analysis of substance use. *Journal of Prevention and Intervention in the Community*, In press.

N. Cliff. *Analyzing multivariate data*. Harcourt Brace Jovanovitch, San Diego, CA, 1987.

C.C. Clogg. Latent class models. In G. Arminger, C.C. Clogg, and M. Sobel, editors, *Handbook of statistical modeling for the social and behavioral sciences*, pages 311–359. Plenum Press, New York, 1995.

C.C. Clogg and L.A. Goodman. Simultaneous latent structure analysis in several groups. *Sociological Methodology*, pages 81–110, 1985.

C.C. Clogg, D.B. Rubin, N. Schenker, B. Schultz, and L. Weidman. Multiple imputation of industry and occupation codes in census public-use samples using Bayesian logistic regression. *Journal of the American Statistical Association*, pages 68–78, 1991.

D.L. Coffman, M.E. Patrick, L.A. Palen, B.L. Rhodes, and A.K. Ventura. Why do high school seniors drink? Implications for a targeted approach to intervention. *Prevention Science*, 8:241–248, 2007.

L.M. Collins. Reliability for static and dynamic categorical latent variables: Developing measurement instruments based on a model of the growth process. In L.M. Collins and A. Sayer, editors, *New methods for the analysis of change*, pages 271–288. American Psychological Association, Washington, DC, 2001.

L.M. Collins. Using latent transition analysis to examine the gateway hypothesis. In D.B. Kandel and M. Chase, editors, *Stages and pathways of drug involvement:*

Examining the gateway hypothesis, pages 254–269. Cambridge University Press, Cambridge, UK, 2002.

L.M. Collins. Analysis of longitudinal data: The integration of theoretical model, temporal design, and statistical model. *Annual Review of Psychology*, 57:505–528, 2006.

L.M. Collins and B.P. Flaherty. Latent class models for longitudinal data. In J.A. Hagenaars and A.L. McCutcheon, editors, *Applied latent class analysis*, pages 287–303. Cambridge University Press, Cambridge, UK, 2002.

L.M. Collins and J.W. Graham. The effect of the timing and spacing of observations in longitudinal studies of tobacco and other drug use: Temporal design considerations. *Drug and Alcohol Dependence*, 68:85–96, 2002.

L.M. Collins and S.E Wugalter. Latent class models for stage-sequential dynamic latent variables. *Multivariate Behavioral Research*, 27:131–157, 1992.

L.M. Collins, J.W. Graham, J.D. Long, and W.B. Hansen. Crossvalidation of latent class models of early substance use onset. *Multivariate Behavioral Research*, 29: 165–183, 1994.

L.M. Collins, J.W. Graham, S.S. Rousculp, and W.B. Hansen. Heavy caffeine use and the beginning of the substance use onset process: An illustration of latent transition analysis. In K. Bryant, M. Windle, and S. West, editors, *The science of prevention: Methodological advances from alcohol and substance abuse research*, pages 79–99. American Psychological Association, Washington, DC, 1997.

L.M. Collins, J.L. Schafer, and C.K. Kam. A comparison of inclusive and restrictive strategies in modern missing-data procedures. *Psychological Methods*, 6:330–351, 2001.

R. Cudeck and M.W. Browne. Cross-validation of covariance structures. *Multivariate Behavioral Research*, 18:147–167, 1983.

C.M. Dayton. Educational applications of latent class analysis. *Measurement and Evaluation in Counseling and Development*, 24:131–141, 1991.

C.M. Dayton and G.B. Macready. Concomitant-variable latent class models. *Journal of the American Statistical Association*, 83:173–178, 1988.

C.M. Dayton and G.B. Macready. Use of categorical and continuous covariates in latent class analysis. In J.A. Hagenaars and A.L. McCutcheon, editors, *Applied latent class analysis*, pages 213–233. Cambridge University Press, Cambridge, UK, 2002.

A.P. Dempster, N.M. Laird, and D.B. Rubin. Maximum likelihood from incomplete data via the EM algorithm. *Journal of the Royal Statistical Society, Series B*, 39: 1–38, 1977.

S.E. Embretson and S.P. Reise. *Item response theory for psychologists*. Lawrence Erlbaum Associates, Mahwah, NJ, 2000.

B.S. Everitt. A Note on Parameter Estimation for Lazarsfeld's Latent Class Model using the EM Algorithm. *Multivariate Behavioral Research*, 19:79–89, 1984.

B.S. Everitt. *The Cambridge dictionary of statistics*. Cambridge University Press, Cambridge, UK, 2006.

H.T. Everson, R.E. Millsap, and C.M. Rodriguez. Isolating gender differences in test anxiety: A confirmatory factor analysis of the Test Anxiety Inventory. *Educational and Psychological Measurement*, 51:243–251, 1991.

A.K. Formann. Linear logistic latent class analysis. *Biometrical Journal*, 24:171–190, 1982.

A.K. Formann. Constrained latent class models: Theory and applications. *British Journal of Mathematical and Statistical Psychology*, 38:87–111, 1985.

E.S. Garrett and S.L. Zeger. Latent class model diagnosis. *Biometrics*, 56:1055–1067, 2000.

A. Gelman and X.L. Meng, editors. *Applied Bayesian modeling and causal inference from incomplete-data perspectives*. Wiley, Hoboken, NJ, 2004.

A. Gelman, X. Meng, and H.S. Stern. Posterior predictive assessment of model fitness via realized discrepancies. *Statistica Sinica*, 6:733–807, 1996.

A. Gelman, J.B. Carlin, H.S. Stern, and D.B. Rubin. *Bayesian data analysis (Second edition)*. CRC Press, Boca Raton, FL, 2003.

W.A. Gibson. Three multivariate models: Factor analysis, latent structure analysis, and latent profile analysis. *Psychometrika*, 24:229–252, 1959.

L.A. Goodman. The analysis of systems of qualitative variables when some of the variables are unobservable. Part I – A modified latent structure approach. *American Journal of Sociology*, 79:1179–1259, 1974a.

L.A. Goodman. Exploratory latent structure analysis using both identifiable and unidentifiable models. *Biometrika*, 61:215–231, 1974b.

L.A. Goodman. On the estimation of parameters in latent structure analysis. *Psychometrika*, 44:123–128, 1979.

L.A. Goodman. Latent class analysis: The empirical study of latent types, latent variables, and latent structures. In J.A. Hagenaars and A.L. McCutcheon, editors, *Applied latent class analysis*, pages 3–55. Cambridge University Press, Cambridge, UK, 2002.

R.L. Gorsuch. *Factor analysis*. Lawrence Erlbaum Associates, Hillsdale, NJ, 1983.

J.W. Graham. Missing data analysis: Making it work in the real world. *Annual Review of Psychology*, 60:549–576, 2009.

J. Guo, L.M. Collins, K.G. Hill, and J.D Hawkins. Developmental pathways to alcohol abuse and dependence in young adulthood. *Journal of Studies on Alcohol*, 61:799–808, 2000.

S.J. Haberman. Log-linear models for frequency tables derived by indirect observation: Maximum likelihood equations. *The Annals of Statistics*, 2:911–924, 1974.

S.J. Haberman. *Analysis of qualitative data: Volume 2. New developments*. Academic Press, New York, 1979.

J.A. Hagenaars. Latent structure models with direct effects between indicators: Local dependence models. *Sociological Methods & Research*, 16:379–405, 1988.

J.A. Hagenaars. *Loglinear models with latent variables*. Sage, Thousand Oaks, CA, 1993.

J.A. Hagenaars. Categorical causal modeling: Latent class analysis and directed log-linear models with latent variables. *Sociological Methods & Research*, 26: 436–486, 1998.

J.A. Hagenaars and A.L. McCutcheon. *Applied latent class analysis*. Cambridge University Press, Cambridge, England, 2002.

G.R. Hancock and K.M. Samuelsen. *Advances in latent variable mixture models*. Information Age Publishing, Charlotte, NC, 2008.

H. Hoijtink. Constrained latent class analysis using the Gibbs sampler and posterior predictive p-values: Applications to educational testing. *Statistica Sinica*, 8:691–712, 1998.

J.L. Horn and J.J. McArdle. A practical and theoretical guide to measurement invariance in aging research. *Experimental Aging Research*, 18:117–144, 1992.

G.H. Huang and K. Bandeen-Roche. Building an identifiable latent class model with covariate effects on underlying and measured variables. *Psychometrika*, 69:5–32, 2004.

P.J. Huber. The behavior of maximum likelihood estimates under nonstandard conditions. In *Proceedings of the Fifth Berkeley Symposium in Mathematical Statistics*, volume 1, pages 221–233, 1967.

K. Humphreys and H. Janson. Latent transition analysis with covariates, nonresponse, summary statistics and diagnostics: Modelling children. *Multivariate Behavioral Research*, 35:89–118, 2000.

J. Jaccard, R. Turrisi, and C.K. Wan. *Interaction effects in multiple regression*. Sage, Thousand Oaks, CA, 2003.

K.M. Jackson, K.J. Sher, and P.K. Wood. Trajectories of concurrent substance use disorders: A developmental, typological approach to comorbidity. *Alcoholism: Clinical and Experimental Research*, 24:902–913, 2000.

L.D. Johnston, J.G. Bachman, and J.E. Schulenberg. *Monitoring the Future national results on adolescent drug use: Overview of key findings, 2004*. National Institutes of Health, Bethesda, MD, 2005.

K. G. Jöreskog and D. Sörbom. *Advances in factor analysis and structural equation modeling*. Abt Books, Cambridge, MA, 1979.

K.S. Kendler, L.J. Eaves, E.E. Walters, M.C. Neale, A.C. Heath, and R.C. Kessler. The identification and validation of distinct depressive syndromes in a population-based sample of female twins. *Archives of General Psychiatry*, 53:391–400, 1996.

R.C. Kessler, K.A. McGonagle, S. Zhao, C.B. Nelson, M. Hughes, S. Eshelman, H.-U. Wittchen, and K.S. Kendler. Lifetime and 12-month prevalence of DSM-III-R psychiatric disorders in the United States: Results from the national comorbidity survey. *Archives of General Psychiatry*, 51:8–19, 1994.

R.C. Kessler, M.B. Stein, and P. Berglund. Social phobia subtypes in the National Comorbidity Survey. *American Journal of Psychiatry*, 155:613–619, 1998.

R. Klein. *Principles and practice of structural equation modeling (Second edition)*. Guilford Press, New York, 2004.

K.J. Koehler. Goodness-of-fit tests for log-linear models in sparse contingency tables. *Journal of the American Statistical Association*, 81:483–493, 1986.

K.J. Koehler and K. Larntz. An empirical investigation of goodness-of-fit statistics for sparse multinomials. *Journal of the American Statistical Association*, 75:336–344, 1980.

R. Langeheine. New developments in latent class theory. In R. Langeheine and J. Rost, editors, *Latent trait and latent class models*, pages 77–108. Plenum Press, New York, 1988.

R. Langeheine. Latent variables Markov models. In A. Von Eye and C.C. Clogg, editors, *Latent variables analysis: Applications for developmental research*, pages 373–395. Sage, Thousand Oaks, CA, 1994.

R. Langeheine and J. Rost. *Latent trait and latent class models*. Plenum Press, New York, 1988.

R. Langeheine and F. van de Pol. A unifying framework for Markov modeling in discrete space and discrete time. *Sociological Methods & Research*, 18:416–441, 1990.

R. Langeheine, E. Stern, and F. van de Pol. State mastery learning: Dynamic models for longitudinal data. *Applied Psychological Measurement*, 18:277–291, 1994.

S.T. Lanza and L.M. Collins. Pubertal timing and the onset of substance use in females during early adolescence. *Prevention Science*, 3:69–82, 2002.

S.T. Lanza and L.M. Collins. A mixture model of discontinuous development in heavy drinking from ages 18 to 30: The role of college enrollment. *Journal of Studies on Alcohol*, 67:552–61, 2006.

S.T. Lanza and L.M. Collins. A new SAS procedure for latent transition analysis: Transitions in dating and sexual behavior. *Developmental Psychology*, 44:446–456, 2008.

S.T. Lanza, B.P. Flaherty, and L.M. Collins. Latent class and latent transition analysis. In J.A. Schinka and W.F. Velicer, editors, *Handbook of psychology, Volume 2*, pages 663–685. Wiley, Hoboken, NJ, 2003.

S.T. Lanza, L.M. Collins, J.L. Schafer, and B.P. Flaherty. Using data augmentation to obtain standard errors and conduct hypothesis tests in latent class and latent transition analysis. *Psychological Methods*, 10:84–100, 2005.

S.T. Lanza, L.M. Collins, D. Lemmon, and J.L. Schafer. PROC LCA: A SAS procedure for latent class analysis. *Structural Equation Modeling*, 14:671–694, 2007.

K. Larntz. Small-sample comparisons of exact levels for chi-squared goodness-of-fit statistics. *Journal of the American Statistical Association*, 73:253–263, 1978.

P.F. Lazarsfeld and N.W. Henry. *Latent structure analysis*. Houghton Mifflin, Boston, 1968.

T.H. Lin and C.M. Dayton. Model selection information criteria for non-nested latent class models. *Journal of Educational and Behavioral Statistics*, 22:249–264, 1997.

R.J. Little and D.B. Rubin. *Statistical analysis of missing data, Second edition.* Wiley, New York, 2002.

E. Loken and P. Molenaar. Categories or continua? The correspondence between mixture models and factor models. In G.R. Hancock and K.M. Samuelsen, editors, *Advances in latent variable mixture models,* pages 277–297. Information Age Publishing, Charlotte, NC, 2008.

F.M. Lord. *Applications of item response theory to practical testing problems.* Lawrence Erlbaum Associates, Hillsdale, NJ, 1980.

G.H. Lubke and B.O. Muthén. Investigating population heterogeneity with factor mixture models. *Psychological Methods,* 10:21–39, 2005.

M.M. Maldonado-Molina, L.M. Collins, S.T. Lanza, G. Prado, R. Ramírez, and G. Canino. Patterns of substance use onset among Hispanics in Puerto Rico and the United States. *Addictive Behaviors,* 32:2432–2437, 2007.

A.L. McCutcheon. *Latent class analysis.* Sage, Thousand Oaks, CA, 1987.

A.L. McCutcheon. Basic concepts and procedures in single- and multiple-group latent class analysis. In J.A. Hagenaars and A.L. McCutcheon, editors, *Applied latent class analysis,* pages 56–88. Cambridge University Press, Cambridge, UK, 2002.

R.P. McDonald. *Factor analysis and related methods.* Lawrence Erlbaum Associates, Hillsdale, NJ, 1985.

G. McLachlan and D. Peel. *Finite mixture models.* Wiley, New York, 2000.

P.E. Meehl. Factors and taxa, traits and types, differences of degree and differences in kind. *Journal of Personality,* 60:117–174, 1992.

W. Meredith. Measurement invariance, factor analysis and factorial invariance. *Psychometrika,* 58:525–543, 1993.

R.E. Millsap. Invariance and measurement and prediction: Their relationship in the single-factor case. *Psychological Methods,* 2:248–260, 1997.

R.E. Millsap. Invariance in measurement and prediction revisited. *Psychometrika,* 72:461–473, 2007.

R.E. Millsap and O.M. Kwok. Evaluating the impact of partial factorial invariance on selection in two populations. *Psychological Methods,* 9:93–115, 2004.

S.L. Morgan and C. Winship. *Counterfactuals and causal inference: Methods and principles for social research.* Cambridge University Press, New York, NY, 2007.

I. Moustaki. A latent trait and a latent class model for mixed observed variables. *British Journal of Mathematical and Statistical Psychology*, 49:313–334, 1996.

B.O. Muthén and L.K. Muthén. The development of heavy drinking and alcohol-related problems from ages 18 to 37 in a U. S. national sample. *Journal of Studies on Alcohol*, 61:290–300, 2000.

B.O. Muthén and K. Shedden. Finite mixture modeling with mixture outcomes using the em algorithm. *Biometrics*, 55:463–469, 1999.

D. Nagin. *Group-Based Modeling of Development*. Harvard University Press, Cambridge, MA, 2005.

K.L. Nylund, T. Asparouhov, and B.O. Muthén. Deciding on the number of classes in latent class analysis and growth mixture modeling: A Monte Carlo simulation study. *Structural Equation Modeling*, 14:535–569, 2007a.

K.L. Nylund, A. Bellmore, A. Nishina, and S. Graham. Subtypes, severity, and structural stability of peer victimization: What does latent class analysis say? *Child Development*, 78:1706–1722, 2007b.

K.L. Nylund, T. Asparouhov, and B.O. Muthén. Deciding on the number of classes in latent class analysis and growth mixture modeling: A Monte Carlo simulation study (Erratum). *Structural Equation Modeling*, 15:182, 2008.

M.E. Patrick, L.M. Collins, E. Smith, L. Caldwell, A. Flisher, and L. Wegner. A prospective longitudinal model of substance use onset among South African adolescents. *Substance Use & Misuse*, 44:647–662, 2009.

B.H. Patterson, C.M. Dayton, and B.I. Graubard. Latent class analysis of complex sample survey data. *Journal of the American Statistical Association*, 97:721–741, 2002.

E.J. Pedhazur and L.P. Schmelkin. *Measurement, design, and analysis: An integrated approach*. Lawrence Erlbaum Associates, Hillsdale, NJ, 1991.

S.F. Posner, L.M. Collins, D. Longshore, and M.D. Anglin. The acquisition and maintenance of safer sexual behaviors among injection drug users. *Substance Use & Misuse*, 31:1995–2015, 1996.

V. Ramaswamy, W.S. DeSarbo, D.J. Reibstein, and W.T. Robinson. An empirical pooling approach for estimating marketing mix elasticities with PIMS data. *Marketing Science*, 12:103–124, 1993.

T.R.C. Read and N.A.C. Cressie. *Goodness-of-fit statistics for discrete multivariate data*. Springer, New York, NY, 1988.

B.A. Reboussin, D.M. Reboussin, K.Y. Liang, and J.C. Anthony. Latent transition modeling of progression of health-risk behavior. *Multivariate Behavioral Research*, 33:457–478, 1998.

B.A. Reboussin, K.Y. Liang, and D.M. Reboussin. Estimating equations for a latent transition model with multiple discrete indicators. *Biometrics*, 55:839–845, 1999.

B.A. Reboussin, E.H. Ip, and M. Wolfson. Locally dependent latent class models with covariates: An application to under-age drinking in the USA. *Journal of the Royal Statistical Society. Series A,(Statistics in Society)*, 171:877–897, 2008.

D. Rindskopf. Linear equality restrictions in regression and loglinear models. *Psychological Bulletin*, 96:597–603, 1984.

D. Rindskopf and W. Rindskopf. The value of latent class analysis in medical diagnosis. *Statistics in Medicine*, 5:21–27, 1986.

J.M. Robins, M.Á. Hernán, and B. Brumback. Marginal structural models and causal inference in epidemiology. *Epidemiology*, 11:550–560, 2000.

D.B. Rubin. Inference and missing data. *Biometrika*, 63:581–592, 1976.

D.B. Rubin. Bayesianly justifiable and relevant frequency calculations for the applied statistician. *The Annals of Statistics*, 12:1151–1172, 1984.

D.B. Rubin. Causal inference using potential outcomes. *Journal of the American Statistical Association*, 100:322–331, 2005.

D.B. Rubin and N. Schenker. Logit-based interval estimation for binomial data using the Jeffreys prior. *Sociological Methodology*, 17:131–144, 1987. ISSN 00811750. URL http://www.jstor.org/stable/271031.

D.B. Rubin and H.S. Stern. Testing in latent class models using a posterior predictive check distribution. In A. Von Eye and C.C. Clogg, editors, *Latent variables analysis: Applications for developmental research*, pages 420–438. Sage, Thousand Oaks, CA, 1994.

J. Ruscio and A.M. Ruscio. Advancing psychological science through the study of latent structure. *Current Directions in Psychological Science*, 17:203–207, 2008.

J. Ruscio, N. Haslam, and A.M. Ruscio. *Introduction to the taxometric method: A practical guide*. Lawrence Erlbaum Associates, Mahwah, NJ, 2006.

J.L. Schafer. *Analysis of incomplete multivariate data*. CRC Press, Boca Raton, FL, 1997.

J.L. Schafer and J.W. Graham. Missing data: Our view of the state of the art. *Psychological Methods*, 7:147–177, 2002.

J.L. Schafer and J. Kang. Average causal effects from nonrandomized studies: A practical guide and simulated example. *Psychological Methods*, 13:279–313, 2008.

G. Schwartz. Estimating the dimension of a model. *The Annals of Statistics*, 6: 461–464, 1978.

L. Sclove. Application of model-selection criteria to some problems in multivariate analysis. *Psychometrika*, 52:333–343, 1987.

A. Skrondal and S. Rabe-Hesketh. *Generalized latent variable modeling: Multilevel, longitudinal, and structural equation models*. Chapman & Hall, Boca Raton, FL, 2004.

M.A. Tanner and W.H. Wong. The calculation of posterior distributions by data augmentation. *Journal of the American Statistical Association*, 82:528–540, 1987.

L.L. Thurstone. An analytic method for simple structure. *Psychometrika*, 19:173–182, 1954.

J.R. Udry. *The National Longitudinal Study of Adolescent Health (Add Health), Waves I & II, 1994-1996; Wave III, 2001-2002 [machine-readable data file and documentation]*. Carolina Population Center, University of North Carolina at Chapel Hill, Chapel Hill, NC, 2003.

J.S. Uebersax. Probit latent class analysis with dichotomous or ordered category measures: Conditional independence/dependence models. *Applied Psychological Measurement*, 23:283–297, 1999.

J.S. Uebersax and W.M. Grove. Latent class analysis of diagnostic agreement. *Statistics in Medicine*, 9:559–572, 1990.

F. van de Pol and J. de Leeuw. A latent Markov model to correct for measurement error. *Sociological Methods & Research*, 15:118–141, 1986.

F. van de Pol and R. Langeheine. Mixed Markov latent class models. *Sociological Methodology*, 20:213–247, 1990.

P.G.M. van der Heijden, J. Dessens, and U. Böckenholt. Estimating the concomitant-variable latent-class model with the EM algorithm. *Journal of Educational and Behavioral Statistics*, 21:215–229, 1996.

W.J. Van der Linden and R.K. Hambleton. *Handbook of modern item response theory*. Springer Verlag, New York, NY, 1997.

R.J. Vandenberg. Toward a further understanding of and improvement in measurement invariance methods and procedures. *Organizational Research Methods*, 5: 139–158, 2002.

R.J. Vandenberg and C.E. Lance. A review and synthesis of the measurement invariance literature: Suggestions, practices, and recommendations for organizational research. *Organizational Research Methods*, 3:4–69, 2000.

W.F. Velicer, A.C. Peacock, and D.N. Jackson. A comparison of component and factor patterns: A Monte Carlo approach. *Multivariate Behavioral Research*, 17: 371–388, 1982.

J.K. Vermunt. Multilevel latent class models. *Sociological Methodology*, 33:213–239, 2003.

J.K. Vermunt and J. Magidson. *Latent GOLD user's guide*. Statistical Innovations, Inc., Belmont, MA, 2000.

J.K. Vermunt and J. Magidson. Latent class cluster analysis. In J.A. Hagenaars and A.L. McCutcheon, editors, *Applied latent class analysis*, pages 89–106. Cambridge University Press, Cambridge, England, 2002.

J.K. Vermunt and J. Magidson. *Technical guide for Latent GOLD 4.0: Basic and advanced*. Statistical Innovations, Inc., Belmont, MA, 2005.

J.K. Vermunt, R. Langeheine, and U. Böckenholt. Discrete-time discrete-state latent Markov models with time-constant and time-varying covariates. *Journal of Educational and Behavioral Statistics*, 24:179 – 207, 1999.

A. Von Eye and C.C. Clogg. *Latent variables analysis: Applications for developmental research*. Sage, Thousand Oaks, CA, 1994.

M. Wedel and W.S. DeSarbo. A mixture likelihood approach for generalized linear models. *Journal of Classification*, 12:21–55, 1995.

H. White. Maximum likelihood estimation of misspecified models. *Econometrica: Journal of the Econometric Society*, pages 1–25, 1982.

K.F. Widaman and S.P. Reise. Exploring the measurement invariance of psychological instruments: Applications in the substance use domain. In K.J. Bryant, M.T. Windle, and S.G. Weat, editors, *The science of prevention: Methodological advances from alcohol and substance abuse research*, pages 281–324. American Psychological Association, Washington, DC, 1997.

K. Yamaguchi. Multinomial logit latent-class regression models: An analysis of the predictors of gender-role attitudes among Japanese women. *American Journal of Sociology*, 105:1702–1740, 2000.

Topic Index

Author Index

Agresti, A., 78, 83, 151, 153, 167, 243
Aiken, L.S., 163
Aitkin, M., 46
Akaike, H., 88
Anderson, D., 46
Anglin, M.D., 223
Anthony, J.C., 225
Asparouhov, T., 9, 108
Auerbach, K.J., 223
Bachman, J.G., 5, 21, 95, 133, 150
Baker, F.B., 6
Bandeen-Roche, K., 46, 107, 149
Bellmore, A., 46
Berglund, P., 4
Bergman, L.R., 8
Bollen, K.A., 6
Box, G.E.P., 82
Bozdogan, H., 88
Browne, M.W., 88
Brumback, B., 159
Bulik, C.M., 4
Bye, B.V., 8
Böckenholt, U., 7, 9, 149, 225
Caldwell, L., 223
Canino, G., 263

Carlin, J.B., 68
Celeux, G., 75
Centers for Disease Control and Prevention, 20, 34
Chung, H., 79, 94, 108, 176, 263
Chung, T., 46
Cleveland, M.J., 176
Cliff, N., 152
Clogg, C.C., 6–7, 22, 113, 147, 171
Coffman, D.L., 4
Collins, L.M., 8, 21, 46, 68, 79, 81, 88, 107–108, 147, 172, 175–176, 182–183, 185–187, 201, 207, 210–211, 221–223, 237, 250, 256, 258, 261, 263
Cressie, N.A.C., 86
Cudeck, R., 88
Curran, P.J., 6
Dayton, C.M., 7, 9, 46, 108, 149
de Leeuw, J., 187
Dempster, A.P., 7, 78
DeSarbo, W.S., 46, 74
Dessens, J., 149
Eaves, L.J., 46
El-Khouri, B.M., 8
Embretson, S.E, 6
Eshelman, S., 4–5

WILEY SERIES IN PROBABILITY AND STATISTICS
ESTABLISHED BY WALTER A. SHEWHART AND SAMUEL S. WILKS

Editors: *David J. Balding, Noel A. C. Cressie, Garrett M. Fitzmaurice, Iain M. Johnstone, Geert Molenberghs, David W. Scott, Adrian F. M. Smith, Ruey S. Tsay, Sanford Weisberg*
Editors Emeriti: *Vic Barnett, J. Stuart Hunter, Jozef L. Teugels*

The *Wiley Series in Probability and Statistics* is well established and authoritative. It covers many topics of current research interest in both pure and applied statistics and probability theory. Written by leading statisticians and institutions, the titles span both state-of-the-art developments in the field and classical methods.

Reflecting the wide range of current research in statistics, the series encompasses applied, methodological and theoretical statistics, ranging from applications and new techniques made possible by advances in computerized practice to rigorous treatment of theoretical approaches.

This series provides essential and invaluable reading for all statisticians, whether in academia, industry, government, or research.

† ABRAHAM and LEDOLTER · Statistical Methods for Forecasting
AGRESTI · Analysis of Ordinal Categorical Data
AGRESTI · An Introduction to Categorical Data Analysis, *Second Edition*
AGRESTI · Categorical Data Analysis, *Second Edition*
ALTMAN, GILL, and McDONALD · Numerical Issues in Statistical Computing for the Social Scientist
AMARATUNGA and CABRERA · Exploration and Analysis of DNA Microarray and Protein Array Data
ANDĚL · Mathematics of Chance
ANDERSON · An Introduction to Multivariate Statistical Analysis, *Third Edition*
* ANDERSON · The Statistical Analysis of Time Series
ANDERSON, AUQUIER, HAUCK, OAKES, VANDAELE, and WEISBERG · Statistical Methods for Comparative Studies
ANDERSON and LOYNES · The Teaching of Practical Statistics
ARMITAGE and DAVID (editors) · Advances in Biometry
ARNOLD, BALAKRISHNAN, and NAGARAJA · Records
* ARTHANARI and DODGE · Mathematical Programming in Statistics
* BAILEY · The Elements of Stochastic Processes with Applications to the Natural Sciences
BALAKRISHNAN and KOUTRAS · Runs and Scans with Applications
BALAKRISHNAN and NG · Precedence-Type Tests and Applications
BARNETT · Comparative Statistical Inference, *Third Edition*
BARNETT · Environmental Statistics
BARNETT and LEWIS · Outliers in Statistical Data, *Third Edition*
BARTOSZYNSKI and NIEWIADOMSKA-BUGAJ · Probability and Statistical Inference
BASILEVSKY · Statistical Factor Analysis and Related Methods: Theory and Applications
BASU and RIGDON · Statistical Methods for the Reliability of Repairable Systems
BATES and WATTS · Nonlinear Regression Analysis and Its Applications
BECHHOFER, SANTNER, and GOLDSMAN · Design and Analysis of Experiments for Statistical Selection, Screening, and Multiple Comparisons

*Now available in a lower priced paperback edition in the Wiley Classics Library.
†Now available in a lower priced paperback edition in the Wiley–Interscience Paperback Series.

BELSLEY · Conditioning Diagnostics: Collinearity and Weak Data in Regression
† BELSLEY, KUH, and WELSCH · Regression Diagnostics: Identifying Influential
  Data and Sources of Collinearity
BENDAT and PIERSOL · Random Data: Analysis and Measurement Procedures,
  Third Edition
BERRY, CHALONER, and GEWEKE · Bayesian Analysis in Statistics and
  Econometrics: Essays in Honor of Arnold Zellner
BERNARDO and SMITH · Bayesian Theory
BHAT and MILLER · Elements of Applied Stochastic Processes, *Third Edition*
BHATTACHARYA and WAYMIRE · Stochastic Processes with Applications
BILLINGSLEY · Convergence of Probability Measures, *Second Edition*
BILLINGSLEY · Probability and Measure, *Third Edition*
BIRKES and DODGE · Alternative Methods of Regression
BISWAS, DATTA, FINE, and SEGAL · Statistical Advances in the Biomedical Sciences:
  Clinical Trials, Epidemiology, Survival Analysis, and Bioinformatics
BLISCHKE AND MURTHY (editors) · Case Studies in Reliability and Maintenance
BLISCHKE AND MURTHY · Reliability: Modeling, Prediction, and Optimization
BLOOMFIELD · Fourier Analysis of Time Series: An Introduction, *Second Edition*
BOLLEN · Structural Equations with Latent Variables
BOLLEN and CURRAN · Latent Curve Models: A Structural Equation Perspective
BOROVKOV · Ergodicity and Stability of Stochastic Processes
BOULEAU · Numerical Methods for Stochastic Processes
BOX · Bayesian Inference in Statistical Analysis
BOX · R. A. Fisher, the Life of a Scientist
BOX and DRAPER · Response Surfaces, Mixtures, and Ridge Analyses, *Second Edition*
* BOX and DRAPER · Evolutionary Operation: A Statistical Method for Process
  Improvement
BOX and FRIENDS · Improving Almost Anything, *Revised Edition*
BOX, HUNTER, and HUNTER · Statistics for Experimenters: Design, Innovation,
  and Discovery, *Second Editon*
BOX, JENKINS, and REINSEL · Time Series Analysis: Forcasting and Control, *Fourth
  Edition*
BOX, LUCEÑO, and PANIAGUA-QUIÑONES · Statistical Control by Monitoring
  and Adjustment, *Second Edition*
BRANDIMARTE · Numerical Methods in Finance: A MATLAB-Based Introduction
† BROWN and HOLLANDER · Statistics: A Biomedical Introduction
BRUNNER, DOMHOF, and LANGER · Nonparametric Analysis of Longitudinal Data in
  Factorial Experiments
BUCKLEW · Large Deviation Techniques in Decision, Simulation, and Estimation
CAIROLI and DALANG · Sequential Stochastic Optimization
CASTILLO, HADI, BALAKRISHNAN, and SARABIA · Extreme Value and Related
  Models with Applications in Engineering and Science
CHAN · Time Series: Applications to Finance
CHARALAMBIDES · Combinatorial Methods in Discrete Distributions
CHATTERJEE and HADI · Regression Analysis by Example, *Fourth Edition*
CHATTERJEE and HADI · Sensitivity Analysis in Linear Regression
CHERNICK · Bootstrap Methods: A Guide for Practitioners and Researchers,
  Second Edition
CHERNICK and FRIIS · Introductory Biostatistics for the Health Sciences
CHILÈS and DELFINER · Geostatistics: Modeling Spatial Uncertainty
CHOW and LIU · Design and Analysis of Clinical Trials: Concepts and Methodologies,
  Second Edition
CLARKE · Linear Models: The Theory and Application of Analysis of Variance

---

FELLER · An Introduction to Probability Theory and Its Applications, Volume I, *Third Edition,* Revised; Volume II, *Second Edition*

FISHER and VAN BELLE · Biostatistics: A Methodology for the Health Sciences

FITZMAURICE, LAIRD, and WARE · Applied Longitudinal Analysis

* FLEISS · The Design and Analysis of Clinical Experiments

FLEISS · Statistical Methods for Rates and Proportions, *Third Edition*

† FLEMING and HARRINGTON · Counting Processes and Survival Analysis

FUJIKOSHI, ULYANOV, and SHIMIZU · Multivariate Statistics: High-Dimensional and Large-Sample Approximations

FULLER · Introduction to Statistical Time Series, *Second Edition*

† FULLER · Measurement Error Models

GALLANT · Nonlinear Statistical Models

GEISSER · Modes of Parametric Statistical Inference

GELMAN and MENG · Applied Bayesian Modeling and Causal Inference from Incomplete-Data Perspectives

GEWEKE · Contemporary Bayesian Econometrics and Statistics

GHOSH, MUKHOPADHYAY, and SEN · Sequential Estimation

GIESBRECHT and GUMPERTZ · Planning, Construction, and Statistical Analysis of Comparative Experiments

GIFI · Nonlinear Multivariate Analysis

GIVENS and HOETING · Computational Statistics

GLASSERMAN and YAO · Monotone Structure in Discrete-Event Systems

GNANADESIKAN · Methods for Statistical Data Analysis of Multivariate Observations, *Second Edition*

GOLDSTEIN and LEWIS · Assessment: Problems, Development, and Statistical Issues

GREENWOOD and NIKULIN · A Guide to Chi-Squared Testing

GROSS, SHORTLE, THOMPSON, and HARRIS · Fundamentals of Queueing Theory, *Fourth Edition*

GROSS, SHORTLE, THOMPSON, and HARRIS · Solutions Manual to Accompany Fundamentals of Queueing Theory, *Fourth Edition*

* HAHN and SHAPIRO · Statistical Models in Engineering

HAHN and MEEKER · Statistical Intervals: A Guide for Practitioners

HALD · A History of Probability and Statistics and their Applications Before 1750

HALD · A History of Mathematical Statistics from 1750 to 1930

† HAMPEL · Robust Statistics: The Approach Based on Influence Functions

HANNAN and DEISTLER · The Statistical Theory of Linear Systems

HARTUNG, KNAPP, and SINHA · Statistical Meta-Analysis with Applications

HEIBERGER · Computation for the Analysis of Designed Experiments

HEDAYAT and SINHA · Design and Inference in Finite Population Sampling

HEDEKER and GIBBONS · Longitudinal Data Analysis

HELLER · MACSYMA for Statisticians

HINKELMANN and KEMPTHORNE · Design and Analysis of Experiments, Volume 1: Introduction to Experimental Design, *Second Edition*

HINKELMANN and KEMPTHORNE · Design and Analysis of Experiments, Volume 2: Advanced Experimental Design

HOAGLIN, MOSTELLER, and TUKEY · Fundamentals of Exploratory Analysis of Variance

* HOAGLIN, MOSTELLER, and TUKEY · Exploring Data Tables, Trends and Shapes

* HOAGLIN, MOSTELLER, and TUKEY · Understanding Robust and Exploratory Data Analysis

HOCHBERG and TAMHANE · Multiple Comparison Procedures

HOCKING · Methods and Applications of Linear Models: Regression and the Analysis of Variance, *Second Edition*

*Now available in a lower priced paperback edition in the Wiley Classics Library.

†Now available in a lower priced paperback edition in the Wiley–Interscience Paperback Series.

*Now available in a lower priced paperback edition in the Wiley Classics Library.

†Now available in a lower priced paperback edition in the Wiley–Interscience Paperback Series.

Regression, *Third Edition*
* MILLER · Survival Analysis, *Second Edition*
MONTGOMERY, JENNINGS, and KULAHCI · Introduction to Time Series Analysis and Forecasting
MONTGOMERY, PECK, and VINING · Introduction to Linear Regression Analysis, *Fourth Edition*
MORGENTHALER and TUKEY · Configural Polysampling: A Route to Practical Robustness
MUIRHEAD · Aspects of Multivariate Statistical Theory
MULLER and STOYAN · Comparison Methods for Stochastic Models and Risks
MURRAY · X-STAT 2.0 Statistical Experimentation, Design Data Analysis, and Nonlinear Optimization
MURTHY, XIE, and JIANG · Weibull Models
MYERS, MONTGOMERY, and ANDERSON-COOK · Response Surface Methodology: Process and Product Optimization Using Designed Experiments, *Third Edition*
MYERS, MONTGOMERY, and VINING · Generalized Linear Models. With Applications in Engineering and the Sciences
† NELSON · Accelerated Testing, Statistical Models, Test Plans, and Data Analyses
† NELSON · Applied Life Data Analysis
NEWMAN · Biostatistical Methods in Epidemiology
OCHI · Applied Probability and Stochastic Processes in Engineering and Physical Sciences
OKABE, BOOTS, SUGIHARA, and CHIU · Spatial Tesselations: Concepts and Applications of Voronoi Diagrams, *Second Edition*
OLIVER and SMITH · Influence Diagrams, Belief Nets and Decision Analysis
PALTA · Quantitative Methods in Population Health: Extensions of Ordinary Regressions
PANJER · Operational Risk: Modeling and Analytics
PANKRATZ · Forecasting with Dynamic Regression Models
PANKRATZ · Forecasting with Univariate Box-Jenkins Models: Concepts and Cases
* PARZEN · Modern Probability Theory and Its Applications
PEÑA, TIAO, and TSAY · A Course in Time Series Analysis
PIANTADOSI · Clinical Trials: A Methodologic Perspective
PORT · Theoretical Probability for Applications
POURAHMADI · Foundations of Time Series Analysis and Prediction Theory
POWELL · Approximate Dynamic Programming: Solving the Curses of Dimensionality
PRESS · Bayesian Statistics: Principles, Models, and Applications
PRESS · Subjective and Objective Bayesian Statistics, *Second Edition*
PRESS and TANUR · The Subjectivity of Scientists and the Bayesian Approach
PUKELSHEIM · Optimal Experimental Design
PURI, VILAPLANA, and WERTZ · New Perspectives in Theoretical and Applied Statistics
† PUTERMAN · Markov Decision Processes: Discrete Stochastic Dynamic Programming
QIU · Image Processing and Jump Regression Analysis
* RAO · Linear Statistical Inference and Its Applications, *Second Edition*
RAUSAND and HØYLAND · System Reliability Theory: Models, Statistical Methods, and Applications, *Second Edition*
RENCHER · Linear Models in Statistics
RENCHER · Methods of Multivariate Analysis, *Second Edition*
RENCHER · Multivariate Statistical Inference with Applications
* RIPLEY · Spatial Statistics
* RIPLEY · Stochastic Simulation
ROBINSON · Practical Strategies for Experimenting
ROHATGI and SALEH · An Introduction to Probability and Statistics, *Second Edition*

*Now available in a lower priced paperback edition in the Wiley Classics Library.
†Now available in a lower priced paperback edition in the Wiley–Interscience Paperback Series.

ROLSKI, SCHMIDLI, SCHMIDT, and TEUGELS · Stochastic Processes for Insurance and Finance

ROSENBERGER and LACHIN · Randomization in Clinical Trials: Theory and Practice

ROSS · Introduction to Probability and Statistics for Engineers and Scientists

ROSSI, ALLENBY, and McCULLOCH · Bayesian Statistics and Marketing

† ROUSSEEUW and LEROY · Robust Regression and Outlier Detection

* RUBIN · Multiple Imputation for Nonresponse in Surveys

RUBINSTEIN and KROESE · Simulation and the Monte Carlo Method, *Second Edition*

RUBINSTEIN and MELAMED · Modern Simulation and Modeling

RYAN · Modern Engineering Statistics

RYAN · Modern Experimental Design

RYAN · Modern Regression Methods, *Second Edition*

RYAN · Statistical Methods for Quality Improvement, *Second Edition*

SALEH · Theory of Preliminary Test and Stein-Type Estimation with Applications

* SCHEFFE · The Analysis of Variance

SCHIMEK · Smoothing and Regression: Approaches, Computation, and Application

SCHOTT · Matrix Analysis for Statistics, *Second Edition*

SCHOUTENS · Levy Processes in Finance: Pricing Financial Derivatives

SCHUSS · Theory and Applications of Stochastic Differential Equations

SCOTT · Multivariate Density Estimation: Theory, Practice, and Visualization

† SEARLE · Linear Models for Unbalanced Data

† SEARLE · Matrix Algebra Useful for Statistics

† SEARLE, CASELLA, and McCULLOCH · Variance Components

SEARLE and WILLETT · Matrix Algebra for Applied Economics

SEBER · A Matrix Handbook For Statisticians

† SEBER · Multivariate Observations

SEBER and LEE · Linear Regression Analysis, *Second Edition*

† SEBER and WILD · Nonlinear Regression

SENNOTT · Stochastic Dynamic Programming and the Control of Queueing Systems

* SERFLING · Approximation Theorems of Mathematical Statistics

SHAFER and VOVK · Probability and Finance: It's Only a Game!

SILVAPULLE and SEN · Constrained Statistical Inference: Inequality, Order, and Shape Restrictions

SMALL and McLEISH · Hilbert Space Methods in Probability and Statistical Inference

SRIVASTAVA · Methods of Multivariate Statistics

STAPLETON · Linear Statistical Models, *Second Edition*

STAPLETON · Models for Probability and Statistical Inference: Theory and Applications

STAUDTE and SHEATHER · Robust Estimation and Testing

STOYAN, KENDALL, and MECKE · Stochastic Geometry and Its Applications, *Second Edition*

STOYAN and STOYAN · Fractals, Random Shapes and Point Fields: Methods of Geometrical Statistics

STREET and BURGESS · The Construction of Optimal Stated Choice Experiments: Theory and Methods

STYAN · The Collected Papers of T. W. Anderson: 1943–1985

SUTTON, ABRAMS, JONES, SHELDON, and SONG · Methods for Meta-Analysis in Medical Research

TAKEZAWA · Introduction to Nonparametric Regression

TAMHANE · Statistical Analysis of Designed Experiments: Theory and Applications

TANAKA · Time Series Analysis: Nonstationary and Noninvertible Distribution Theory

THOMPSON · Empirical Model Building

THOMPSON · Sampling, *Second Edition*

THOMPSON · Simulation: A Modeler's Approach

THOMPSON and SEBER · Adaptive Sampling

*Now available in a lower priced paperback edition in the Wiley Classics Library.

†Now available in a lower priced paperback edition in the Wiley–Interscience Paperback Series.

THOMPSON, WILLIAMS, and FINDLAY · Models for Investors in Real World Markets
TIAO, BISGAARD, HILL, PEÑA, and STIGLER (editors) · Box on Quality and
    Discovery: with Design, Control, and Robustness
TIERNEY · LISP-STAT: An Object-Oriented Environment for Statistical Computing
    and Dynamic Graphics
TSAY · Analysis of Financial Time Series, *Second Edition*
UPTON and FINGLETON · Spatial Data Analysis by Example, Volume II:
    Categorical and Directional Data
† VAN BELLE · Statistical Rules of Thumb, *Second Edition*
VAN BELLE, FISHER, HEAGERTY, and LUMLEY · Biostatistics: A Methodology for
    the Health Sciences, *Second Edition*
VESTRUP · The Theory of Measures and Integration
VIDAKOVIC · Statistical Modeling by Wavelets
VINOD and REAGLE · Preparing for the Worst: Incorporating Downside Risk in Stock
    Market Investments
WALLER and GOTWAY · Applied Spatial Statistics for Public Health Data
WEERAHANDI · Generalized Inference in Repeated Measures: Exact Methods in
    MANOVA and Mixed Models
WEISBERG · Applied Linear Regression, *Third Edition*
WELSH · Aspects of Statistical Inference
WESTFALL and YOUNG · Resampling-Based Multiple Testing: Examples and
    Methods for p-Value Adjustment
WHITTAKER · Graphical Models in Applied Multivariate Statistics
WINKER · Optimization Heuristics in Economics: Applications of Threshold Accepting
WONNACOTT and WONNACOTT · Econometrics, *Second Edition*
WOODING · Planning Pharmaceutical Clinical Trials: Basic Statistical Principles
WOODWORTH · Biostatistics: A Bayesian Introduction
WOOLSON and CLARKE · Statistical Methods for the Analysis of Biomedical Data,
    Second Edition
WU and HAMADA · Experiments: Planning, Analysis, and Parameter Design
    Optimization, *Second Edition*
WU and ZHANG · Nonparametric Regression Methods for Longitudinal Data Analysis
YANG · The Construction Theory of Denumerable Markov Processes
YOUNG, VALERO-MORA, and FRIENDLY · Visual Statistics: Seeing Data with
    Dynamic Interactive Graphics
ZACKS · Stage-Wise Adaptive Designs
ZELTERMAN · Discrete Distributions—Applications in the Health Sciences
* ZELLNER · An Introduction to Bayesian Inference in Econometrics
ZHOU, OBUCHOWSKI, and McCLISH · Statistical Methods in Diagnostic Medicine

---

*Now available in a lower priced paperback edition in the Wiley Classics Library.
†Now available in a lower priced paperback edition in the Wiley–Interscience Paperback Series.

CPSIA information can be obtained
at www.ICGtesting.com
Printed in the USA
BVHW082122050619
550306BV00011B/125/P

9 780470 228395